Paul Frankel:
Common Carrier of Common Sense

Paul Frankel: Common Carrier of Common Sense

A Selection of his Writings, 1946–1988

EDITED BY
IAN SKEET

Published by Oxford University Press
for the Oxford Institute for Energy Studies
1989

Oxford University Press, Walton Street, Oxford OX2 6DP

Oxford New York Toronto
Delhi Bombay Calcutta Madras Karachi
Petaling Jaya Singapore Hong Kong Tokyo
Nairobi Dar es Salaam Cape Town
Melbourne Auckland

and associated companies in
Beirut Berlin Ibadan Nicosia

Oxford is a trade mark of Oxford University Press

OIES books are distributed in the United
States and Canada by PennWell Books, Tulsa, Oklahoma

© Oxford Institute for Energy Studies
1989

British Library Cataloguing in Publication Data
Frankel, Paul
    Paul Frankel: common carrier of common sense:
    a selection of his writings, 1946–1988
    1. Petroleum industries
    I. Title   II. Skeet, Ian   III. Mabro, Robert
    338.2'7282

ISBN 0-19-730009-X

Typeset by Oxford Computer Typesetting
Printed in Great Britain
by Billing & Sons Ltd., Worcester

The *Oxford Institute for Energy Studies* was established in 1982 to foster through research and advanced study the mutual understanding of the various parties involved in international energy. The Institute has a publication programme whose aim is to disseminate the results of research undertaken by its staff and associates, selected papers presented at sessions of the Oxford Energy Seminar, and the works of eminent specialists, industrialists and statesmen who have made important contributions to the energy debate.

The Institute is pleased to publish in this spirit this anthology of the works of Paul Frankel. We would like to thank the following for their kind co-operation in allowing us to include material previously published by them in this collection: BBC, *Middle East Economic Survey*, *Petroleum Intelligence Weekly*, *Petroleum Review*, *Petroleum Times*, Platt's Oil Price Service. We would also like to thank Ted White, Peter Regnier and Ken Budge of the PEL Group for their assistance in this project.

Contents

Abbreviations

AFRA	Average Freight Rate Assessment
AG	Aktiengesellschaft (Joint-stock Company)
AGIP	Azienda Generale Italiana Petroli
AIOC	Anglo-Iranian Oil Corporation
ARAMCO	Arabian–American Oil Company
BBC	British Broadcasting Corporation
BHP	Broken Hill Proprietary
BNOC	British National Oil Corporation
BP	British Petroleum
BRP	Bureau de Recherches de Pétrole
BTU	British thermal unit
CAMPSA	Compañía Arrendataria del Monopolio de Petróleos, S.A.
CEI	Center for Education in International Management
CFP	Compagnie Française des Pétroles
CIEC	Conference on International Economic Cooperation
DCF	Discounted cash flow
DDT	Dichloro-diphenyl-trichloro-ethane
DEA	Deutsche Erdöl Aktiengesellschaft
ERAP	Entreprise de Recherches et d'Activités Pétrolières
ENEL	Ente Nazionale per L'Energia Elettrica
ENI	Ente Nazionale Idrocarburi
f.o.b.	Free on board
ICI	Imperial Chemical Industries plc
IEA	International Energy Agency
IMI	International Management Institute
IPAC	Iran Pan American Oil Company
IPC	Iraq Petroleum Company
IRI	Istituto per la Ricostruzione Industriale
KNPC	Kuwait National Petroleum Company
LDCs	Less developed countries

MEES	*Middle East Economic Survey*
MIT	Massachusetts Institute of Technology
NIOC	National Iranian Oil Corporation
OECD	Organisation for Economic Co-operation and Development
ÖMV	Österreichische Mineralölverwaltung AG
OPEC	Organization of the Petroleum Exporting Countries
PEL	Petroleum Economics Ltd
PIW	*Petroleum Intelligence Weekly*
SIRIP	Société Irano-Italienne des Pétroles
TAPline	Trans-Arabian Pipeline
TBA	Tyres, batteries and accessories
TNEC	Temporary National Economic Committee (USA)
UN	United Nations

Introduction

Paul Frankel was born in Vienna in 1903. He studied there, attended Vienna University and obtained a Doctorate in Political Economy in 1926. He joined an independent oil refiner and marketer dealing mainly in paraffin wax and lubricants and moved to Poland in 1927. In 1929 he transferred to Danzig and remained there until, in 1937, he moved to England. He joined Manchester Oil Refineries, whose founders and directors were his compatriots Franz Kind and Georg Tugendhat. He himself became a director of the company after the war and resigned on the death of Kind in 1955. In the same year he established Petroleum Economics Limited. He was Chairman, then President, of PEL until 1987 when he was appointed Life President of the PEL Group. On his eighty-fifth birthday, in late 1988, he has decided formally to withdraw from the oil industry with which he has been associated for over sixty years. This is, therefore, a suitable occasion on which to take a backward look at Paul Frankel's achievements and his contribution to the industry. This has been done by selecting a number of his writings over the years and attaching to each a brief introductory and linking remark. But, first, there are a number of things to be said about his career.

The period up to 1955 was, as it turned out, a preparation for PEL. Theoretical economics in Vienna was followed by the practical experience of a tough marketing job; being a special products trader in Danzig for Polish exports was no sinecure. Then, the accident of the war years provided him with the basics of journalism and the opportunity for some more rigorous analysis. The immediate result of this varied experience was *Essentials of Petroleum*, published shortly after the end of the war, in 1946. This made him a name. It brought him the opportunity to carry on with his broadcasting and writing role which had started in the war. He began to be known within the industry and to know the leaders of the industry. He was able to travel to the USA; he already, of course, knew Europe better than most.

All this led to an important turning-point in his life. In the early 1950s the major oil companies – roughly, the Seven Sisters, as they could then still be legitimately described – were anxious to establish an independent research institution. It was to be called the International Bureau for Research in Oil Economics or some equivalent title. Paul Frankel was approached to create and direct it. This was an appealing and tempting offer. He was on the point of accepting when he realized – rightly, one suspects, given the context and the time – that he would never in practice be independent, but that he would, whatever the theory, find himself bounden to those who financed the institution. So he refused. Shortly after, Kind died, and Frankel resigned from Manchester Oil Refineries to set up his own consultancy company, PEL.

He started PEL with his own limited, but by now increasing, reputation, an assistant Walter Newton (who would remain with PEL for nearly thirty years and be its Managing Director for ten), very little capital and three clients. The three clients were ENI, CFP and Deutsche Erdöl (later replaced by Veba); they have been clients ever since. He made an early decision never to take on the majors as clients (whether they had wished it or not) on the grounds that they would tend to swamp him. He believed that the non-majors needed him and that he could help them in a way that he could not help the majors. He never changed his mind or his policy on this point.

By any standards PEL has been a successful venture. Paul Frankel can be justifiably proud of his achievement as a consultant. He has, moreover, managed to do what so few of his contemporaries in this line of business have achieved; he has passed on his firm to another generation – Ted White, who joined PEL in 1964 and became Chairman of the PEL Group in 1986, and his team. So often the enterprise has died when the founder has retired or become uninterested – or failed to retire.

PEL will always be associated with the name of Frankel, but he will be just as satisfied to be remembered by the many who have listened to him speak in seminars, conferences or courses as an expositor, or educator. He is, in a real sense, far more of a teacher than a business man and the activities of teaching – but on a commercial, not a charitable, basis – have been an important element in his career. Today there are many courses covering (or at least purporting to cover) the economics of the oil industry but Frankel really invented them. Northwestern University in Illinois gave him his first opportunity in the early 1960s, but soon after he started his own, under the auspices of PEL and CEI (now IMI) in Geneva. The PEL Seminar – the Frankel Seminar – started in 1969 and shows no sign of ending. He has consistently supported and assisted the Institute of Petroleum in a similar endeavour.

Educating the oil industry to understand itself is one thing; another is to encourage policy-makers to do the things you feel they should do, or at least to see things the way you think they should. This too has been a continuing preoccupation of Paul Frankel's. In the *Essentials* he first analysed the role that he saw as essential for government (and this will recur in the pages of this volume) and it has never been far from his thoughts, his writings or his activities. It is, of course, a frustrating business, since the process of persuading politicians to listen, to understand and finally to act on what seems self-evidently rational is seldom rewarding. Frankel, however, has never stopped trying.

Two relatively high-profile attempts were through the Zurich Group in 1972 and the Frankel Group in 1983. The former sought to create an alternative to the concession basis of activity by the major consortia in the Middle East in the context of OPEC militancy towards the existing concession holders. Frankel would say today that there was not enough intellectual structure to the proposals, but they were anyway overtaken by events. The Frankel Group was, perhaps, a more serious effort to influence policy in the context of OPEC disarray, falling prices, reducing oil demand and the threat of fall-off of investment. Two or three papers were written during 1983–5, the need for co-operation and dialogue between producers and consumers stressed, and a positive governmental or intergovernmental initiative underlined. Nothing came of this, the timing was wrong, the psychology at fault. The point, however, was that Frankel believed he should do something and did it.

In a less formal way, of course, he was 'doing something' in most of his publications, meetings and contacts. In this sense he was a lobbyist as well as consultant, educator and business man. In *Oil: The Facts of Life* he coined the phrase 'the Common Carrier of Common Sense' as descriptive of the kind of person who might be the catalyst for rational governmental action. This has seemed a suitable epitaphic title to be attached to Paul Frankel and has, therefore, been used for this volume. Slightly stilted perhaps, slightly dated, but this seems appropriate in signalling how he has straddled the decades.

In looking back and trying to tie together Paul Frankel's career and achievements it is easy enough for us to remember what he has done, what he has taught. Part of the purpose of this volume is to help the next generation or two to be equally aware of this. The choice of pieces to include has necessarily been somewhat arbitrary, particularly as I have only had access to part of his large output. It has been based on two requirements. The first was to deal adequately with the economics of the industry as seen by Frankel – the principles expounded in the *Essentials* and developed from there. The second was to illustrate how he responded at the time to some of the more important occurrences in the oil market. It was theoretically possible to divide up the result by subject or to treat it chronologically. I chose the latter since it seems to me to provide a thread of coherence which might otherwise be lacking. I hope that my own short introductory linking remarks will assist this process. I should add that the last piece, Postscript, had not been written at the time the selection was made but I was aware that it would be written and it seemed only fair and appropriate to give the last word to Paul Frankel himself.

The other purpose of the volume is to record the moment at which Frankel has decided finally to retire and to express our gratitude for his participation in and contribution to the oil industry. Of the various honours he has received over the years the one of which I believe him to have been most proud was to be awarded the Cadman Medal. That was a symbol of recognition by his professional colleagues that he was one of them and that it was not necessary to belong to one of the large companies to be so recognized.

Perhaps one of the reasons why the establishment found it easy to accept him was that Paul Frankel was – happily, still is – an internationally minded person. Oil companies tend to be multinational. Paul Frankel is essentially European. 'I was born in Austria', he says 'but I do not feel an Austrian. I became British, but one cannot become English. If there's no Europe, I have no fatherland.'

Ian Skeet
Oxford, June 1988

1   Essentials of Petroleum: A Key to Oil Economics

The war years, although he was busy enough with broadcasting and other work, provided Paul Frankel with time in which to ponder the underlying economic principles on which the oil industry was based and in terms of which it functioned. He was able to analyse its structure and to reach conclusions on the nature of competition in the industry. In 1946 he published Essentials of Petroleum, *which became something in the nature of a textbook for those in and joining the industry.*

Frankel himself has often said, only half-jokingly, that he has spent the rest of his career restating the principles that he first formulated in 1946. They have, indeed, stood up staunchly to the passage of time. Investment remains the key to the industry; the relationship of fixed to variable costs is unchanged; units are large; the industry is not self-adjusting in terms of supply/demand/price. One element has changed – the idea that demand is impervious to price, but this only became apparent post-1980 when crude oil prices had been multiplied by ten in ten years. The low price elasticity of oil products remained a principle borne out in practice for at least thirty years after the publication of Essentials. *Perhaps one other change of detail is worth noting. While exploration was in 1946 the important cost (and risk) factor, and production a lesser investment imposition, today the position has been reversed. The success of exploration remains uncertain, of course, but cost is concentrated in the production phase.*

Another of Frankel's principles has been under some investigation. The compulsion towards integration has seemed to some observers to have disintegrated under the post-1973 changes in the industry. On this matter, however, the case is still open, no verdict has been given. What seems to have happened is that the virtues of integration have not been debased and are still, for the most part, sought by those who think they can be achieved.

Finally, there is the question of direction and supervison of the oil industry – 'oil is too serious an affair to be left to oilmen'. Nothing

Chapman & Hall, London, 1946; 2nd edition, Frank Cass & Co. Ltd, London, 1969.

*that Frankel says on this subject is out of date even if the solutions –
rather, the compromises reached – may still seem imperfect to him. It is a
debate that constantly recurs throughout this volume, as it has in the
thinking and career of Frankel himself.*

It would be impractical to repeat the whole of the Essentials *here.
Fortunately Peter Regnier, today a director of the PEL Group, made an
abridged edition of the book some years ago (it was never published or
printed) and I have used this to make a further abridgement for the
purposes of this volume. Regnier made it clear that in his abridgement he
had altered nothing in the light of hindsight. Nor have I. Much of
what has been removed is notes on the original text and quotations from
other sources. What is left is, as Regnier said, the 'Essentials of the*
Essentials*', but for those who may be dissatisfied there is always the
remedy of returning to the original.*

PART I

ECONOMICS OF A LIQUID

The *leitmotif* of any discussion about petroleum must be its
liquid state.

The problems involved will be considered first against their
technical, or rather scientific, background; the fact that most
petroleum products are volatile liquids delimits their possibilities and determines their role among similar or competing
materials.

Starting with this knowledge, the specific features of the
exploration of oilfields and the exploitation of oil-wells will
have to be investigated.

The next stage is refining. Here the fact that a liquid cannot
be 'handled' – in the original sense of the word – fixes the
pattern of the industry. Refining requires little labour but
elaborate plant.

Lastly, we shall see the consequences of the liquid state of
petroleum in transport and marketing where it entails the use
of specialized equipment which puts the oil trade into a category all of its own.

We shall discover that there exist certain traits which

permeate through the whole of the oil industry, and only by appreciating their common denominators can we understand properly how vital it is to think always in terms of the *whole* industry rather than to try to solve the problems of any one of its component parts as if it were self-contained.

Crude Production

The governing factor in the economics of crude oil production is that exploration and drilling are expensive, while the actual cost of lifting the oil from the subsoil, that is, the cost of exploitation, is relatively low. In other words, *capital investment is necessarily heavy, whereas current expenses are light.* This proportion of fixed and variable cost provides a set of circumstances characteristic not only of the production side of the industry, but also of some of its later phases.

The Job of Finding Oil. In order to appreciate the problems and difficulties of locating oil and getting at it, we must bear in mind that it is to be found in porous rock at various levels deep under the surface and that, in spite of all the knowledge and experience gained over almost a century, the occurrence of oil in profitable quantities at any given point cannot be deduced theoretically, but can ultimately be proved only by the act of bringing down a bore. This obviously leads to the drilling of a great many holes which give no tangible result, which are 'dry'.

This is one of the reasons why crude production has always been beset by the problem of overheads. In the long run, the productive wells have to 'carry' those that are unsuccessful. Since the greater part of costs is preliminary to actual production, it is imperative that the operator extracts the maximum from a producing well, because his cost per barrel decreases rapidly as output increases. Owing to the unorthodox character of crude production, the aggregate cost of achieving production is seldom borne in mind and, therefore, not always recovered. The occasional stupendous fortunes derived from a lucky strike boost the hopes of would-be investors; the many failures are forgotten. In Adam Smith's words: 'Man is an incorrigible optimist. He despises future risks and underinsured trusts blindly in his star.'

But, even without taking the heavy cost of exploration and of dry holes into account, it is still true that the drilling cost of any individual well is out of all proportion to the expense of getting oil to the surface once the productive stratum is reached.

When, some years ago, Myron W. Watkins enumerated 'the three distinguishing characteristics of crude petroleum',[1] he mentioned first 'exhaustibility', to which I do not pay as much attention as he did, since it is a feature shared by all mineral resources, though in different degrees. The other two characteristics were 'concealment' and 'fugacity', that is, 'its fugitive character arising from its fluidity'.

So far I have only referred to 'concealment' – that is, to the difficulty of locating the material and of getting at it – which gives such a decided twist to the economics of producing crude. But Watkins' last characteristic, the 'fugacity' of crude, is of equal importance; indeed, it has received more attention than any other like problem. That the oilman's interest should have focused on 'its fugitive character arising from its fluidity', is probably due to the fact that 'concealment' is a factor in the *struggle of Man against Nature*, whereas 'fugacity' is a part of the *competitive contest between Man and Man*, and, consequently, more exciting.

Law of Capture. In the United States, as in many other countries, the rights to subsoil resources generally belong to the owners of the land and, as the boundaries of several properties may cut across what is popularly called an 'oil pool' – and, more correctly, a rigid sponge or an area of permeable sands – there may arise the problem of an interplay of several interests drawing upon one common source of supply. As fluids find their own level, and as any crude which is within a certain structure of porous rock tends to migrate towards the point of lowest pressure, the way in which one owner works his 'lease' will inevitably affect the interests of all the others. The most common solution of this problem, an automatic one as it were, is to leave it to the competing parties to fight it out. This

---

[1] Myron W. Watkins, *Oil: Stabilization or Conservation?*, New York and London, 1937, p. 29.

conception was based on the 'law of capture', according to which a wild beast is deemed to be the property of the owner of the land on which it is slain or intercepted. Whatever repercussions this law may have had in the realm of hunting, as far as oil is concerned it has, of necessity, led to a rapid development of the resource. It has now, for very good reasons, become fashionable to scoff at such methods of rapid development, but, sound as the arguments against them are, it is doubtful whether the 'oil age', i.e. the swift progress of the internal combustion engine, could have come off without the unrelenting pressure exercised by an almost too plentiful supply of cheap oil. It may well be that these somewhat primitive methods were exactly what was required to make the young industry, in the first instance, aggressive and, in due course, great.

Time Is *Money*. The main factors the crude producers had to consider, in the circumstances, were:

(1) The necessity of quick production to make his heavy investment pay as soon as possible.
(2) The commitments towards the property owner who was to receive a royalty, and who usually granted the lease on condition that it should be exploited within a specified time.
(3) The danger of 'his' oil being drained away by his neighbours.

All three points make for swift action, especially in the case of operators with limited resources. But the third point is the most potent of all when you remember that this type of producer is usually operating in fields where property holdings are in many hands.

Such circumstances brought about the system of 'offset wells'; the man who developed a lease tended to start drilling near its boundaries, and this forced his neighbour in turn to forestall him by doing the same on the other side.

These tactics led to cramped conditions and to a spacing of wells according to other than purely technical considerations. But the real failing of this system is the effort wasted in drilling more holes than necessary and the loss of underground gas pressure caused by opening too many 'valves' at any one time.

There can be little doubt that operators in subdivided pools pay for high initial yields by impairing seriously ultimate crude oil recovery.

Conservation. The case for considering any one pool as a 'unit' whose exploitation is to run on communal lines, giving each holder of a lease an 'undivided interest in the entire area', is a very impressive one.

The justification for curbing the rights attached to the individual lease by unitization and the still more sweeping principle of *proration*, i.e. of limiting the 'allowable' output, lies in the belief that uncontrolled exploitation spells 'waste'. It must be clearly understood that people mean quite different things by waste; some refer to the reduction of the aggregate quantity of oil and of gas obtained from a pool by inadequate production methods – 'technical' waste, for short – while others mean 'commercial' waste, due to the production of more oil than can be disposed of at economical or remunerative prices, with the after-effect of a possible shortage at a later date.

It was not until the stupendous glut in 1927, followed by renewed 'over-production' scares in the early thirties, that a powerful section of the American oil industry attempted to tackle the problem seriously, and managed to enlist the support of the Roosevelt Administration. The fact that the industry awoke to its long-term interests only when its short-term business prospects were threatened may be a reflection upon the acumen of the majority of its leaders, but it does not affect the value of the principles involved. The wisdom of 'planning' the operations of extractive industries, so as to avoid violent oscillations of stocks and prices, may be a debatable point, and in respect of 'economic waste' the oil industry is on the same footing as others, but, if we visualize economic waste against the background of its technical equivalent, we are bound to appreciate that *both together* are somewhat formidable.

If industries, whose raw materials are in unlimited supply and in which exploitation methods do not necessarily imperil the future, elect to put their trust in day-to-day expediency reflected in the workings of a free market, that is their business. But if ever the case for co-ordination of interests has been made, it is on the producing side of the oil industry.

Refining

Cost: Fixed and Variable. The character of an industry is to a great extent determined by the relation of its variable to its fixed costs. At one extreme is the entrepreneur who 'gives out' to outworkers who use their own tools; his manufacturing costs are variable since they consist entirely of wages paid to casual piece-workers. At the other extreme there would be an imaginary plant – automatic, requiring no attention, and having a negligible fuel consumption.

The first type of industrialist can, provided there are suitable applicants for his type of job, hire and fire at will; he can adjust output instantly and completely to meet changing market conditions because all his costs are variable. The owner of the automaton, however, has only 'fixed' cost and, even if the plant is shut down or run at reduced throughput, he must still allow for interest and depreciation at almost the same rate as when it is operating to capacity.

This does not mean that high fixed cost makes for bad business as compared with 'light' industry. But it does mean that the policy of an industry with heavy investment and low variable cost must differ from one in which wages and power are the biggest items. The industry in which cost rises and falls according to output will, generally speaking, show considerable elasticity, that is, it will contract and expand easily to adjust itself to the state of the market. If the bulk of your cost is directly attributable to actual production, if it is so-called *prime cost* you will obviously tend to adjust your output without delay to the amount you think you can dispose of at a price which, at least, covers this cost.

If, on the other hand, the greater part of your expenditure is *fixed cost* which is incurred independent of output, the prime cost of each individual article produced is low and prices can fall a good deal below the level at which *all* costs are covered and real profits are made, before it pays to reduce the rate of production. Such industry is therefore less elastic in so far as – once the investment has been made – the incidence of fixed cost makes full utilization of the plant imperative and production can easily get out of hand. If your fixed cost (interest, depreciation, maintenance, and administration) is N, it affects each of 100 units produced to the extent of $N/100$; if, *with the same equipment*, you can produce 150 units the cost per unit will not be quite as low as $N/150$, since not all costs are fixed, but it will be sufficiently near this figure to induce you to strive for maximum output within the limits of your plant capacity. In other words, the cost per unit goes down rapidly as production is stepped up. It follows that once the plant is built, it is difficult to keep its throughput down.

Therefore, in 'light' industries supply will tend to follow smoothly the fluctuations of demand whereas industries requiring 'heavy' investments will work spasmodically, either outpacing demand or falling behind it. (One of the possible causes of inadequate supply is the necessity of being sure of a market before large amounts are invested in a plant which can only be run economically at full capacity.)

Students of the oil industry can have no difficulty in deciding into which category it falls. We have seen how much the cost of exploration and drilling outweighs actual lifting cost in production, and this ratio obtains, though in a different way, at the refining stage.

Labour in Oil Refining. Labour is of high quality, but, as far as numbers and the total of wages are concerned, it is of limited importance and does not really influence financial results. Certainly its traditionally excellent labour relations do credit to oil refining all over the world, but this happy state of affairs is not, as is sometimes supposed, due only to the virtue and public spirit of the industry's leaders. It nearly always pays to treat a skilled and specialized man well, but it is much easier for an industry whose wages are a minor item to live up to this principle than for one which its wage bill makes or breaks. The central problem of refining economics is what has been called the 'gap between the prime cost and the total cost'.[2] The former is very low, indeed more than half of the total cost

---

[2] E. A. G. Robinson, *The Structure of Competitive Industry*, London and Cambridge, 1937 (Cambridge Economic Handbooks-VII. General Editor: D. H. Robertson), p. 94.

is in overheads and cannot be directly related to the amount produced. The famous 'last 10 per cent' of a refiner's through-put which involves next to nothing in cost apart from chemicals, and can be sold at any old price without making the accountants blush, is firewood for kindling price-war conflagrations. Once a refinery is built its owners are prisoners in the hands of their investment, none of the emergency exits of other industries are open to them; if it comes to the point either a refiner will manage to win through by fighting for a market which will absorb his full-scale production, or he will perish, unless, of course, there is some arrangement which will compensate low throughput by securing high prices.

Full Employment of Plant. Briefly, refining is a matter of equipment and the success of a plant depends on whether proper use can be made of it or not. It must work to a stable programme and as nearly as possible to capacity.

The process of production in a modern refinery is, by its very nature, continuous. This did not obtain to such an extent as long as the original methods of batch treatment prevailed both on the distilling and on the refining side, but pipe still, cracking and solvent refining plants are designed to run for a long stretch, only shutting down for cleaning and maintenance. As it is entirely unsatisfactory to start up and close down any of these plants at frequent intervals – on account of heat economy and off-grade products derived at the beginning and at the end of each run – the alternatives to working at much less than the rated capacity are: continuous operation at reduced throughput or plant shut-down for a period of months which involves the availability of sizeable storage facilities to tide regular business over the recurrent shut-down periods.

The former method is appropriate for moderate deviations from rated capacities; a reduction of 5 or 10 per cent is of little importance, but the curve of cost per unit rises steeply thereafter, because by this method practically *all* items of expenses down to fuel cost remain unaltered, and the plant manager soon reaches the point where total, if temporary, closing down is the better way out. This applies particularly when, by nature of the plant, corrosion of vessels and damage to boiler pipes, etc. is less when the plant does not operate.

In either case, however, the refiner is faced with the necessity of providing not only for interest and depreciation, but also for practically all his personnel. Most operatives in a refinery are specialists steeped in the technicalities of their jobs and difficult to replace. A refiner will much rather hang on to them than try to save their wages in a shut-down period. They are, in fact, an integral part of the plant. They have attained the rank of an 'overhead'.

The position is aggravated still further by the peculiar function of the distillation unit within the framework of a refinery. In a textile factory, the spinner or weaver who works a number of units in parallel can carry on with only some of them working. But in a refinery – and in a steelworks – it is the main unit, the lifeline as it were, that is affected by fluctuations in throughput. Distillation plants and blast furnaces can neither be by-passed nor can they, for technical reasons, be broken down to the capacity of any one of the several treating and finishing plants which rely on the primary stage for their intake of intermediate products.

The refinery also relies on reasonably balanced outlets for quite different products which, especially before the advent of cracking and hydrogenation, were anything but readily interchangeable. It is, however, the superiority of large-scale plant, and not only the working to capacity, which is such a cardinal feature of the refining industry. It is not only the usual policy of 'safety in numbers', nor the familiar advantage of doing something in a big way; it is due to the fact that the development of refinery technique – the general progress from the era of topping plants through that of thermal cracking to that of complete refineries incorporating catalytic cracking, reforming and other plants – has put a premium on refining in big, compact units.

All these features point in the same direction; they show petroleum refining hurried by unavoidable technical forces towards working to capacity and concentration in big units. The ensuing competitive position is one so strained that relief must be sought in one way or the other. The alternatives are a kind of struggle for the survival of the fittest, who eventually *controls* the market, or co-operation of the competitors with a view to *regulating* the market.

Transport and Marketing

In a book on the British Gas Industry[3] Philip Chantler points out that its structure would have been entirely different had the individual consumer been able to buy the gas 'ex works' and to take it away in containers. The fact that such a form of delivery was impossible, and that an elaborate system of underground pipes was required to solve the specific problem of carrying gas to the consumer created the set of circumstances which resulted in the establishment of local monopoly and, as its concomitant, public regulation.

Specialized Equipment. In the same way the job of transporting a liquid is 'an integral part of its production', and it has always been of vital importance for the oil industry as a whole. At the user end it is the same story, and boiler fuel oil for ships is a case in point: its ascendancy over coal is partly due to its capacity to flow right into the burner with no need for shovelling and stoking. The very process of using the fuel is a transport function.

The history of petroleum would have been different if the material to be burnt in lamps or in internal-combustion engines could have been made into a powder, packeted, and sold in general stores. Lubricating oil, the one liquid petroleum product used in comparatively small quantities and the least inflammable of the range, is a borderline case and, when packed in small containers, does not provide a particular transport and marketing problem. Solid products like paraffin wax seem to belong to a different world altogether. Crude oil, however, and the principal products are handled in large quantities, 'in bulk', and there *the main consequence of the liquid state of petroleum is that it requires specialized equipment.*

Solid substances usually share means of overland transport and places of storage with other materials. The same truck or lorry can carry coal, steel or timber and the same warehouse can accommodate all sorts of raw materials or finished products. Large-scale transport of a liquid, however, can only be effected in tank wagons or pipe-lines. These are specially

[3] *The British Gas Industry*, An Economic Study, by Philip Chantler. Manchester, 1938, p. 66.

designed for a liquid – the same applies to tank installations – and are thus no good for anything else.

The consequences are:

(1) Whereas other trades can rely upon means of transport catering for a host of materials and are thus not compelled to provide machinery of their own, the oil industry has always had to consider transport as being a major problem to be solved within its own orbit.

(2) This being the case, transport is not, as it is for some other industries, an accidental item. It is a constituent factor which has considerably influenced the structure of the industry. As a matter of fact, the development of oil economics can best be described in terms of transport: first, it was the epoch of the barrel (a barrel is still the unit of measurement for crude and heavy oils), then came the era of bulk containers on their own wheels and, as consumption gets bigger and more concentrated, so the pipe-line age draws nearer.

(3) We are faced with some of the familiar features of crude production and refining.

Where to Store Crude. Before and after each transport stage materials have to be stored; indeed, storage is but a phase of the transport function.

Storage of huge quantities of crude surging from a newly-struck gusher used to be a difficult problem but then it was realized that the most appropriate place for storing crude was the subsoil: the 'conservation' idea is based on the conception that oil should be withdrawn at a controlled rate, and could best be left where it was against the day when it was actually required. Such a procedure, however, hinges on unified control of a field; oil can be stored underground only if it belongs to a 'Common Pool'. That, to meet difficulties, one can go one step further still in such a 'storage policy', is demonstrated by the practice of pumping unwanted fractions back into wells rather than wasting them or putting up cumbrous storage tanks.

Shore Installations. In areas contiguous to oilfields the problem of storing large quantities at the user end does not arise, since supply can be effected by a series of consecutive deliveries in rail or road tank wagons or continuously by pipe-line. Where, on the other hand, long-distance transport necessitates shipment by river, coastal or deep-sea tanker, the problem arises of how to accommodate the oil prior to loading and after unloading. The obvious shortcomings of its most primitive solution – that of filling from and unloading into rail cars, not to mention the use of drums at either end, which has been tried from time to time – prove the inescapable necessity of being able to use permanent shore or river-side installations.

The specialized character of equipment suitable for handling liquids has not usually attracted people who concentrate on transportation of miscellaneous goods, and the oil industry has had to run the whole show itself. When and where this happened business came to be controlled by those who could muster a turnover sufficiently big and constant to justify and support the machinery and organization involved. Only at focal points of trade and traffic was it feasible to develop 'neutral' storage installations, on public wharfinger lines, catering for a considerable number of individual traders, big and small.

Before the war the Scandinavian countries and Finland provided an excellent example of the former type: the almost complete control of these markets by Major Companies was mainly due to the fact that they, and they alone, controlled bulk shore installations capable of holding full cargoes. Less wealthy independent interests were hampered by a vicious circle: they could not acquire a clientele whose requirements would have been sufficient to vindicate investment in shore installations without first providing the installations.

Market 'Units'. Storage facilities have also played a vital role in a quite different sphere of oil marketing, in that of 'retail outlets'. Few features are of greater importance for the structure of a market than its 'units'; the smaller they are the more diversified will the market be.

In the early days of the oil trade the 'unit' was the barrel, since actual delivery to the retailers was made in casks. Whoever had sufficient turnover to sell the contents of a barrel within a reasonable time could, subject to certain fire prevention regulations, enter the field as a retail outlet.

Long before the motor car era, however, the first steps had been taken to devise storage and delivery methods which made proper allowance for the fact that it was a liquid that was to be marketed. It has often been said that Standard Oil owed their paramount success to their exploiting the advantages of being able to use pipe-lines for long-distance transport of crude oil, but the consequences of another development, on a more modest scale, have not generally been appreciated. Apart from their refining advantages, Standard owed their leading position in the kerosine market to their method of supplying the finished products to the customer. They appear to have been the first – not only in the States but in most European countries – to devise a method of storing kerosine immediately prior to its sale to the consumer. They provided shops with a little tank from which the retailer could draw the required quantity at the time of sale, and they replenished the stock in the tank from horse-drawn road tank wagons. The retailer was never faced with the necessity of handling packages, he was only concerned with the liquid itself. Such an arrangement offered considerable advantages, and retailers were easily persuaded to sign an agreement to the effect that they would not sell any kerosine except that supplied by the company providing the tank.

Once again it can be seen that success of commercial and industrial policies mainly depends upon their initiators making use of certain natural factors, upon their harnessing the tides. It was the result of a shrewd and correct appraisal of a technical problem by Rockefeller, and one of the factors of complete and lasting success, that Standard Oil carried the liquid in bulk and deposited it in semi-permanent containers at the retail end rather than shifting to and fro smallish containers which are difficult to handle on the spot, and thus are not satisfactory as storage receptacles.

These are facts of the greatest theoretical significance: *the liquid state of petroleum entails the tendency to develop 'units' of a higher order.* As soon as a certain turnover is reached the barrel system becomes outclassed by *a complex organization whose unit is not,* as one would perhaps be inclined to assume, *the individual retail storage tank, but the whole machinery of bulk terminals and road tank wagons without which it could not function.*

The most recent example of transport and storage arrangements being the controlling factor on the marketing side is the scramble for outlets connected with airports. It is obvious that the marketing 'unit' of aviation engine fuel is a big one, and this prevents competition below a certain size getting anywhere near it.

Links in the Chain. The fact that oilmen have always had to look after transport functions themselves has given a fillip to vertical integration. That the producer has to deliver the crude to the refiner, or that the latter has to fetch it from the fields, helps to break down the dividing line between these two phases of the industry, and the same holds good in respect of refining and marketing.

The importance of transport is still further increased by the fact that oil has to be transported *twice:* from the well to the refinery, from the refinery to the consumer. This applies to many industries, but here it has a peculiar implication because the volume of the material to be shifted is about the same in either case. Iron ore plus coke are much bulkier than steel, the volume and weight of aluminium is only a fraction of that of bauxite. Thus transport matters less for the finished article. But a modern oil refinery turns out almost as much in products as it takes in in the form of crude. These facts have caused the most efficient refineries to be built as near as possible to consuming centres, even if this involved a long haul from the oilfields. It was easier to make use of adequate transport methods for one product – crude – moved in quantities as big as all the several finished products together.

The 'Empties'. One of the greatest snags encountered in transporting a liquid is the need for returning containers empty before they can be used again. This is one more consequence of the self-contained character of the transport of a liquid. Most other vehicles can be used for some other job at or near the place where they are discharged. With the exception of traffic radiating from some sources of raw materials, means of conveyance can be used both ways. The charges for long-distance passenger traffic are based on similar assumptions, and taxicab fares in urban areas would have to be much higher were the drivers not allowed to pick up passengers *en route*.

The cost of the return of empties – of barrels and drums, road and rail tank wagons, river barges, and deep-sea tankers – is considerable, much bigger than is usually realized. The nominal fee railways charge for this service do not cover the cost involved, but it cannot fail to be taken into consideration when the rates for the trip with a 'payload' are worked out.

A Perfect Carrier. Pipe-lines are the only means of transport of a liquid or a gas not involving the return of the empty container or its elimination after one trip. Nothing but the material itself is moved, the pipes which 'contain' it are themselves stationary and thus allow of continuous operation.

The pipe-line, given a certain set of conditions, is the perfect carrier of liquids, and also one of the most ancient. We know of bamboo pipe-lines in the Cathay of Marco Polo's times, and we need only look at the number of aqueducts built by the Romans to realize how great and how obvious the advantages of piping water must have been ever since the art of engineering advanced to the construction of such lines.

In searching for the fundamental reason for the superiority of a line of pipes over a number of containers, we find that, by filling the liquid into a barrel or a tank wagon, which after all is nothing but an outsize drum, we solve the transport problem by putting a protective skin round the liquid and by actually carrying a solid, i.e. the container. At the stages of filling and emptying we take the liquid character of the 'fare' into account, but not at the stage of actual transport.

In the case of a pipe-line, however, we do not eschew the fact that our material is a liquid; on the contrary, we take advantage of its capacity of flowing and of offering comparatively little resistance to changes of form. *Indeed, the essential difference between pipe-lines and all other means of transport is that by the latter the oil can be carried from one place to the other* in spite of *its being a liquid, in the pipe-line because it is one.*

An Integral Part. We saw that the necessity of building and maintaining elaborate and specialized storage installations had an integrating and concentrating effect; the same applies to an even higher degree to pipe-lines. As pipe-line transportation is economically feasible only if there is a continuous flow on a considerable scale, discrimination is inevitable against those competitors whose turnover or financial resources do not allow them to build a pipe-line. This factor was fully realized in the United States by about 1910, when the status of a Common Carrier was imposed on company-run pipe-lines, making them bound to accept any oil tendered to them.

To '*divorce*' pipe-lines from the industry does not seem to be in keeping with the technique and the traditions of the pipe-line system. It would be more appropriate to acknowledge fully the deep integration of the oil transport system in the industry itself, and to use its controlling position for co-ordinating and adjusting developments within the industry as a whole. This problem will loom large when 'policies for the industry' are discussed in the concluding chapter.

PART II

PRICE STRUCTURE

The Influence of Demand on Price

The price of a commodity depends, as far as demand is concerned, on what the buyer can do with it and whether he can do without it. The supplier, on the other hand, must consider the cost of production and alternative uses for raw materials and for capital and labour employed. Variations in any of these elements which determine the market price will affect *all* the others, and whether they will go a long way or be checked at an early stage is dependent on the 'response', as it were, with which they meet.

What does, in fact, go to make up the price of petroleum, and how are its fluctuations to be explained? The price of motor spirit, to take the premier product first, is, as far as demand is concerned, determined chiefly by the service it renders in the engine, and by the relative importance of road transport in a given area as opposed to competing means of conveyance; another factor is technical suitability and availability of alternative fuels – for instance, diesel oil and synthetic products – for internal-combustion engines.

Elasticity of Demand. The best way to understand the mechanism of pricing is to find out what happens if a commodity becomes dearer or cheaper than it was before. According to the text-books, the opportunities for selling a commodity increase if the price goes down, because it comes within the range of buyers who could not afford or would not buy it at the previous price level, and in some cases those who bought it before will increase their purchases if more of the commodity is to be had for the same amount of money. The opposite happens in the case of a price increase – *the higher the price the narrower the market.* Economists have suggested that price movements are to a great extent self-regulating: an increase in price, due to demand exceeding supply, leads to a reduction of sales which will, in due course, restore the balance of supply and demand; a decrease in the price of a commodity will boost its sales and the 'surplus' supply which may have caused the previous price level to give way will readily be taken up by an expanding market.

The tendency to react in such a manner to price fluctuations has been called 'elasticity', and the more elastic a market is the less will be its price fluctuations.

Natural silk is a characteristic example of a commodity with a highly elastic price structure. At a fancy price it was for hundreds of years available only to the wealthiest; even when large quantities became available, during the last century or so, its price only came down very gradually – never did the bottom fall out of the market. This was due to the fact that each of the subsequent reductions of price, *however small*, opened up vast new sales opportunities among people who had always been willing, but were only now able to avail themselves of an article of the inherent excellence of silk.

Motor Spirit Demand Not Price Elastic. Neither of these automatic checks appears to exist in the case of petrol. Its price can, within reason, be increased without anything like a commen-

surable drop in sales and, on the other hand, even the most
radical reduction of price will fail to make markets expand
proportionately unless it is accompanied by other developments which have nothing to do with petrol.

In fact, motor spirit is a perfect example – as, indeed, to an
even greater degree is lubricating oil – of an 'auxiliary' commodity: they are used so that other goods can be put to use.

The running of a private car involves fixed and variable
costs. The former includes the price of the car, the cost of a
garage, the annual licence fee and also the cost of insurance.
Compared with these expenses which accrue independent of
the mileage of the vehicle, the variable costs – mainly engine
fuel, lubricants, and tyres – are relatively small. Once a man
has bought a car and has paid all the expenses that go with
ownership it does not make sense to cut down its use, for only
by using it well and truly can he justify all the expenditure he
has incurred so far. This does not imply that the price of fuel
has no influence whatever on the amount used.

Lubricants: Less Elastic Still. This feature is still more recogniz-
able in the market for lubricating oil, and its very peculiar
character is due to the great value of the lubricated engine or
plant compared with the trifling cost of even the most expensive oil. In the motor spirit market, where it is somewhat
difficult to make a serious case for preferring any particular
brand or type on the grounds of its superior quality, price is to
a great extent determined by competition, whose nature will
be described presently, but lubricants are for very good
reasons less competitive.

The nature of lubrication was little understood for a long
time, and the relative merits of individual oils were difficult to
assess. This unscientific approach to lubrication led to the
vague idea, seldom based on anything more than superstition,
that some oils had a certain something the others hadn't got.
So the lube oil market became a happy hunting ground for
those who knew how to exploit the anxiety of a user who,
when choosing the oil for his engine or machinery, had constantly to bear in mind that a considerable investment was at
stake. In this way certain manufacturers – who mostly, it is
true, had a suitable oil to sell – built up a reputation which

enabled them *to appropriate a substantial part of the big difference between what a good oil costs to make and the remarkable damage a bad oil can cause.*

Indeed, the prices of some of the more famous brands of engine, turbine or transformer oils contain elements which are in the nature of witch doctor's fees, combined with some sort of insurance premium, and cannot be considered purely as rewards for the supply of goods. This description of the lube oil set-up is not intended to suggest that the makers of branded oils were or are making unjustified or extortionate profits, but the colossal margin between production cost and the price which can be obtained for a trusted oil has led to marketing methods of almost unparalleled extravagance. Advertising, high commissions to distributors, fancy sales' aids, and comprehensive service systems – to say nothing of downright bribery – swallow up a great part of the money the consumer pays. All this, however much it may blur the picture, does not affect the underlying fact that, once the strictly competitive character of an 'auxiliary' commodity is removed, monopolistic prices can easily be maintained at practically any level.

Tax Gatherer's Paradise. To return to motor spirit: here the tax collector assumes the role of those who, in the sphere of lubricants, took advantage of their opportunities. Gasoline taxes have been such a success everywhere, because they answer all the requirements of efficient direct taxation. Owing to the widespread use of the commodity it yields a magnificent return, and it does not kill the goose that lays the golden eggs.

Other taxes, like those imposed on tobacco, spirits, or entertainment, are successful because a great part of the population regards these luxuries almost as necessities, and is prepared to sacrifice most other pleasures in order to be able to smoke, to drink and to go to the cinema. Petrol is very nearly in the same category, except in so far as its low elasticity for a rise in price is paralleled by an equally low elasticity for a fall. What happens if its price falls is due to the 'auxiliary' character of petrol. If the price of beer or cigarettes were halved, it would probably result in an immediate and considerable increase in consumption, but this would not be the case with petrol.

Would people who were already running a car be likely to double their mileage?

No Serious Competitors. The second factor that is likely to limit variations in the prices of commodities, that is, the possibility of gaining ground from a competitive material by a decrease and the danger of seeing one's markets invaded by newcomers in the case of an increase, is, in the short run at least, almost non-existent for petroleum products. With the exception of kerosine in the early days and fuel oil, within certain limits, they have never taken over existing markets; they have created new markets of their own, and where they compete with other materials price is not the point at issue, but rather the strength of a superior performance by a given product. Road transport, the principal outlet for petroleum products in spite of all other developments, does not score because it does a job cheaper than could some other form of transport, but because it does a job which, in present conditions, no other type of transport could do.

One of the reasons why petroleum products have a confined, and thus sheltered, market is because, as a liquid, it is comparatively easy to transport. The fact that the main potential competitor for petroleum used as fuel is coal, a solid, makes it awkward to switch over from one to the other without altering a good deal of equipment, and this makes such markets un-reactive so far as short-term price fluctuations are concerned.

While it is true that there are no other materials which, in their own way, could give similar service, it is also true that there are no equivalent materials to take the place of petro-leum products in the internal-combustion engine or in the field of lubrication.

Considering all the factors I have enumerated there can be little doubt that on the demand side there exist few of the acknowledged automatic safeguards against rapid and exten-sive fluctuations in prices, which can easily be driven sky-high or knocked down to a fraction of their previous level without much relief from an expanding or narrowing market. It now remains only to review price structure from the supply angle.

Factors on the Supply Side

From what I have said of the advantages of large-scale production and of the paramount importance of working to capacity, it will follow that increase in output will, in the long run, tend to lead to price reductions. This, however, applies without qualification only to marketing and refining, not to production; crude is an irreplaceable, i.e. diminishing, resource, and it is by no means impossible that, at a given moment, increase in demand will make it necessary to resort to more expensive production methods, and also that crude, no longer being discovered at an appropriate rate, will assume a scarcity value which it has not as yet.

Elasticity of Supply. We have seen that high prices need not curb demand, but it is equally true that low 'unremunerative' prices will sometimes fail to keep output in check. The small producer will – as will his opposite numbers among rubber or coffee planters – try, if prices are low, to produce more and not less, so as to make a bare living, and the highly capitalized refiners and marketers may find it more profitable to maintain their turnover and, incidentally, their standing in the industry, even if this involves heavy losses.

This point, in particular, is fundamental to any history of the petroleum industry, and there will be more to be said about it; here and now, however, one factor matters which, incidentally, cuts right across all that has been said about how little demand is affected by the price level of petroleum products.

Whereas it may not make much difference to the consumer, it is a matter of life and death for a distributor to have to sell at a penny more than do his competitors of equal rank. This does not mean that a smaller local firm could not sell at a somewhat lower price than the big combines without disturbing the market unduly, but none of the big firms with their nationwide sales organization could afford, for any length of time, to be undersold by a competitor of equal standing.

The consumer-in-the-street is faced with a number of brands which he considers as being, generally speaking, on one and the same quality level. If, then, the price of one of the competitors is lowered the others must either follow suit or face a big and immediate loss in gallonage.

The combined weight of all these features exerts a considerable strain on the price-forming machinery, at least as far as the main products – motor spirit, kerosine, gas oil, and fuel oil – are concerned. The low elasticity of demand extends no hope of relief by expansion or contraction of outlets; heavy fixed costs preclude an elastic production policy and, finally, the extremely competitive character of the products we have just discussed makes it all the more difficult to smooth over differences when they arise. Indeed, it can be said that in the absence of automatic appliances for its prevention, the smallest spark is liable to develop into a general conflagration.

There are, however, redeeming constituents in the economics of petroleum which deserve our closest attention: the fact that out of a given crude can be made a whole range of finished products, which can be sold at different prices; and, secondly, the possibility of varying the percentage yield of these various products which has endowed the petroleum industry with that degree of freedom essential to its perfection and prosperity.

Shifting Borderlines. Crude oil, this mixture of chemical compounds so complicated that it still defies definition, could not be used in its original state; for all practical purposes it was, on the one hand, too heavy, dark, and smelly, and, on the other, too inflammable. The problem of selective segregation was solved by evaporating the crude and by condensing it in fractions – in fact, by distillation. The dividing line between the fractions being purely arbitrary, there is considerable latitude as far as actual yields are concerned. There has always been a kind of no-man's land – or both-men's land, if you like – between, say, motor spirit and kerosine, and also between the latter and gas oil, which could be considered as belonging to one or the other of the neighbouring fractions according to what the refiner was after.

When the problem of supplying sufficient light motor fuels by traditional methods became unmanageable – there would then have been hardly enough crude to go round on the basis of an average of 20 to 25 per cent yield of motor spirit obtained by straight-run distillation – the cracking technique, heat treatment of heavy oil under pressure, was developed. Thus the average yield of light fractions was raised to more than 45 per cent, and residual oils, otherwise just fuel to be burnt under the boiler, were transformed into high-grade material. Two birds were killed with one stone, since without cracking, even if there had been enough crude, the quantity of heavier oils which would have had to be made to get the necessary amount of petrol would have created a glut in the market, and might have brought down the price to next to nothing.

As it was, however, the stability of the market was assured by this very opportunity to vary the amount of the several products originating from a given crude. It was also open to operators to develop such crudes as yielded most of the products which were in demand at the time. On a limited scale this made for a fairly *high elasticity of supply*.

By-Products All. The relation of prices to yields can only be fully understood if we keep in mind that *any one petroleum product can be made only if others are derived simultaneously.*

The occurrence of what is called 'joint products', which cannot be produced one without the other, is not peculiar to petroleum – agriculture provides a great number of examples of the wool/mutton type – but for petroleum the interplay of the several products is of a very special significance if only because there is a whole gamut of them.

There is, first, the difficulty of assessing the cost of any one product. The refiner knows what it costs him to make the whole lot together, and it is possible in some cases to single out certain finishing processes, like fractionation of spirits and chemical treatment of lubes, but it is practically impossible to find out how much of running expenses and overheads should be allocated to any one product.

Ever since the early days of petroleum have people racked their brains to find some sort of makeshift solution of this insoluble problem. The obvious idea has always been to pick out the most important product – kerosine in the early days and subsequently gasoline – and to consider all the others as by-products. To this way of thinking the premier product, as I have called it, would be held to bear the brunt of manufacturing costs, but there would be continuous adjustments by transferring part of the cost to such by-products as could bear it by virtue of the prices they could fetch in the markets for which they catered.

The shift of importance from one product to another which goes on continuously justifies the idea that each type can be looked upon as being a by-product of all the others or, better still, that all of them are to be considered as *co-products*. As it is impossible to work out – except for finishing processes – the production cost of the individual products, we shall find that their actual prices are determined by a method which runs parallel to that of costing just mentioned: *each product sells at the price its market will bear.*

Discrimination. Any student of economics is familiar with this conception as the one on which railway tariffs are based and, indeed, the similarity of underlying principles is striking. Originally the idea of railway freight rate-making was to charge freight according to weight or volume carried, but it was soon discovered that such an average rate was prohibitively high for certain bulky and cheap materials, whereas other high-priced light-weight goods could have paid much more without their users feeling the pinch. This is one reason for what has been called freight discrimination; the other arose from the fact that railways are the classical example of undertakings with heavy fixed cost, whereas their prime cost – especially for 'additional' traffic beyond a certain minimum – is extremely low. Such relation of fixed to variable cost always creates the temptation to charge high rates for traffic which is secure, and to cut rates in cases where there is danger of competition.

Criticism against such apparent injustice was soon silenced, because it was clear to anybody who cared to investigate that railways could not operate on a flat rate, and the same principle was in due course accepted for public utility undertakings such as gas and electricity works, whose fixed/variable cost ratio is of a similar order.

The fact that railways appear to supply *one* commodity, the 'ton/mile', electricity undertakings the 'unit', and gas works the 'therm', at varying prices to different customers made the fact of discrimination obvious. In the case of petroleum products it was disguised because the products themselves were different. This difference is, however, as we have seen, more apparent that real, the whole petroleum range consisting of co-products which are dependent upon each other's production.

The low price of certain products is dependent on the sale of one or more 'premium' materials. Originally kerosine – 'illuminating oil' (or just 'petroleum refined'), as it was then called – was the mainstay with all others being grouped under the heading 'By-Products'. Later, the price for 'naphtha and gasoline' rose above that of kerosine. Subsequently the gap between gasoline, on the one hand, kerosine and gas oil on the other, became considerable, but the advent of cracking, of the diesel engine, and of agricultural uses of kerosine tended to equalize the opportunities of the three with the almost immediate result that their prices drew closer together and gasoline became noticeably cheaper.

The different prices of petroleum products and their continuous variation are therefore not 'just a ramp', but are the one and only method of keeping the industry on an even keel. Once again we can see the problem of oil against the background of other industries (including railways) with which it has some basic features in common.

It may be argued, however, that discrimination in the strict sense of the word was possible only within a monopoly, as it exists for various reasons in the realm of railways and public utilities. If there obtains a state of monopoly where a seller's price policy is not affected by competition, then the absence of suitable substitutes for most petroleum products and the low elasticity of demand for some of them create a semi-monopoly for the oil industry as a whole. It is – with the exception perhaps of fuel oils – possible to supply the several markets at different prices and remain in *each* particular case on a lower price level than potential competitors. This is, on the grand scale, tantamount to the railways' policy of encouraging 'additional' freight at the expense of the 'safe' one.

There still remains the question of how competition *among* refiners, which qualifies the statement on the 'monopoly' of the industry, affects the picture *as a whole*. It certainly does so to a great extent in so far as every refiner inclines to concentrate on products which pay and to cut out dead wood. Such competition would appear to eliminate the possibility of charging prices for certain products up to the limit of what the market can bear. On the other hand, this is mitigated by the fact that, however flexible the production process may be, there are definite limits to the shifting of fractions, limits which are set by the chemical structure of a given crude oil, and, incidentally, by the cost of the conversion processes involved.

PART III

THE SHAPE OF THE INDUSTRY

If I had to sum up the results of my investigation and to define, in as few words as possible, the basic feature of the petroleum industry, I should say that what matters most is *that it is not self-adjusting*. Everything so far has pointed in the same direction:

The aleatory character of drilling coupled with high exploration cost and low cost of exploitation;
the unwieldy relation of fixed and variable cost in refining, transport, and marketing; and, finally,
a price structure that allows for ups and downs which fail to bring relief from dearth or glut.

All these facts make for continuous crises: 'the problem of oil is that there is always too much or too little'.[4] Hectic prosperity is followed all too swiftly by complete collapse, and redress can be hoped for only from the efforts of 'eveners', adjusters and organizers, whose success derives from the very peril to which the industry must succumb if they were not to lay down the law. If this is the layout of the industry – what is its history?

[1] Myron W. Watkins, *Oil: Stabilization or Conservation?*, New York and London, 1937, p. 40.

The Great Plan of John D. Rockefeller

Within a few years of Drake's discovery of oil in commercial quantities, two main trends could be distinguished: it was evident that oil was being produced at a rate which left consumption far behind and yet, at the same time, when oil was actually running to waste, people began to worry about the impending exhaustion of the resource.

Of the two threats, of that of 'too much' and 'not enough', the former was the more immediate and the more consistent (but the latter, the spectre of eventual shortage, remained as an undertone and sprang into prominence immediately when, for a time, no big new pools were discovered).

Already early in 1869 – years before the first attempt at oil monopoly was made – a Petroleum Producers' Association cropped up in the Pennsylvanian Oil Region 'to protect the interests of the well owners'. From then onwards time and again the same problem arose: how to keep production of new wells within the limits set by actual demand. No reader who has followed me in my survey of the basic facts of the industry can be in any doubt why there was such a strong urge towards an understanding between competing producers right in the heyday of *laisser faire*. Such understandings were not, in the first instance, designed with the deliberate purpose of charging customers extortionate prices, indeed, when things took a turn for the better they swiftly disappeared. They were nothing but emergency measures, applied only when the bottom had fallen out of the market.

The elimination of the weaker competitors by the mechanism of price fluctuations, acceptable in markets where the mills grind slowly, was always felt to be inadequate in the realm of crude production where 'jerky' developments were unavoidable. People who live on the banks of a torrent are bound to think of building dykes.

Early Essays in Restriction. The first endeavour to unite oil producers on a strictly voluntary basis, however, failed, as did all subsequent schemes of a similar character. If we consider for a moment the structure of any such association, we shall realize why this fate was inevitable. The immediate incentive for an understanding among competing producers was usually the discovery of a new and – at least in its early days – highly prolific field which upset the existing market situation in a complete, albeit temporary, fashion. The clamour for the only possible relief, for a limitation of new drilling and in the last resort, for a partial shut-down of producing wells, would first come from owners of older wells whose production was already on the down grade, and who could less easily afford to accept low prices than those whose flush production helps to make up for a small return per barrel. But this is, after all, only a difference in the degree of concern about low prices. Once the problem is settled, as to how the sacrifices, which the 'understandings' involve, are to be shared, the scheme may become operative.

Smaller supply will send prices up to a satisfactory level. The very success of such a scheme, however, is always liable to be its undoing: the better the price the greater is the temptation to increase one's output by overstepping the allotted 'quota', and those who cheat first actually reap the double benefit of good prices *and* large sales. Once this habit spreads, and a contagious disease it is, it will be only a short time before those who have not benefited from such practices will smash the agreement altogether. Such stories of the making and breaking of horizontal agreements are not, of course, confined to the oil industry. However, the dynamic character of petroleum, unpredictable discoveries, and difficulties of storage, have rendered these problems particularly acute. Their solution is at once vital and difficult, absolutely imperative and almost impossible. Although there is more to be said on this point, I only want to suggest here that such an agreement between a great many independent operators who reserve their right to withdraw from it at any moment is nothing but a temporary expedient. Its structure has much in common with the ancient Polish diet where the dissent of any one member – the *liberum veto* – could throw the whole machine out of gear. The weakness of such an understanding is that while it needs the consent of an overwhelming majority to launch it, it can be wrecked by the defection of a small, if determined, minority.

Such a system is not as democratic as it at first appears.

Democracy works only if the minority accepts the ruling of the majority, because both have certain values in common. An association of straight competitors with equal opportunities may have its day, but it will not last, since the interests of the participants are, however paradoxical it may seem, to such an extent *identical* that they *cannot*, in the long run, be *compatible*.

The urgent need for bringing supply in line with demand, and the great difficulties for producers in achieving this end, were well understood. It was Rockefeller who, when asked by an investigating committee if his monopoly of oil refining and oil transportation had not prevented the producer from getting his full share of the profits, once said:

> The dear people, if they had produced less oil than they wanted, would have got their full price; no combination in the world could have prevented that, if they had produced less oil than the world required.[5]

The competitive position of crude producers was poised more delicately still by the fact that – in the circumstances as they prevailed in the United States – the number of well owners was unavoidably greater than that of their only customers, the refiners. It is thus not surprising that the first effective and properly thought-out attempt to obtain control of the industry was born and bred in the refining sphere. Whereas, especially under primitive conditions, success in the drilling of wells was a matter of daring and luck, the qualities which made a good refiner were very different. From him technical ability and accomplishment in matters of organization were vital. These features were, however, not alone sufficient to lift those who possessed them above the rank and file. Refining, though a narrower field than production, was still too widespread to be 'controlled'. As a matter of fact, none of the usual features which help to establish monopolistic control over a trade was present in the refining industry of those days; raw material was not scarce, nor was it in the hands of a few producers; patented processes or manufacturing secrets did not exist, nor was the capital required for building an efficient refinery excessive. Only if a bottleneck could be discovered or

---

[5] Ida M. Tarbell, *The History of the Standard Oil Company*, London, 1905, Vol. II, p. 157.

created, would there be a real opportunity to regulate the industry.

Control of Keypoints. What does control of an industry mean? Is it necessary to own all its resources and equipment to exercise control? Obviously not, as such complete sway could be achieved only *after* a great measure of control had been established. What control presupposes is effective power in certain parts of the industry which, by reason of their role within its structure, carry without much further effort all the rest of it. The task of destroying Germany's war potential by bombing, first undertaken in earnest in 1943, could not destroy each and every aircraft or engine works. So the Allies set out to eliminate ball-bearing factories, of which there were comparatively few, in order to smash one vital link without which most of the others were of little use. Concentration of attacks on synthetic oil plants and, earlier on, on marshalling yards was the result of the same conception – such is the fine art of 'control'.

First Bottleneck: Rail Transport. Strangely enough, the key to mastery over the petroleum industry lay outside its own orbit. It was to be offered on a silver plate by the railways competing for oil freight. Thirty years later, when the public began to wonder how certain groups had managed to become all-powerful, it was difficult to realize to what extent practices which seemed downright unfair in 1905 were an integral part of the set-up of the seventies. At that time the competition between the several railways serving any one area was at its peak. As they had not achieved co-ordination of their services and tariff structure – a goal to be reached only some time later – they concentrated on securing the freight of the more potent shippers by offering them certain advantages, whereas the smaller fry had to pay the 'official' rates. This practice was by no means confined to oil traffic, it just so happened that its repercussions were particularly severe in the petroleum industry, and there only because of the importance of the transport factor to the industry, discussed so fully earlier in this book.

In their own way the railways acted rationally: once it was realized – as realized it had to be – that discrimination between various types of cargoes was unavoidable, there was no reason why they should not also discriminate in favour of the larger customers whose patronage would help them a long way on the road to prosperity. Lesser concerns had no bargaining power, the volume of their freight was too small to matter much either way: they were the hindmost and the devil took them. Moreover, the big shippers were not only the bone of contention between competing lines, they were equally important, nay indispensable, once the railways had concluded one of their frequent, if short-lived, 'understandings' on freight and tariffs.

This weapon of preferential freights has been wielded with great success and it helped towards establishing what, in due course, became the *de facto* monopoly of the 'Standard' – but it is as well to appreciate that *this monopoly was the direct outcome of its antithesis, the violent competition amongst the railways.*

Monopoly in the Making. Although developments in the American oil industry during the period from 1870 to 1910 were dominated by the personality and the conception of John D. Rockefeller and his Standard Oil Company, it is pretty certain that he was not the first to envisage the possibilities of establishing 'control' of the oil industry as a whole. The authors of the first attempt at using transport as the Archimedean fulcrum wherefrom the whole industry could be levered at will, are not known, but if we can rely on Ida Tarbell's description, they must have had a pretty shrewd idea of its possibilities:

> In the fall of 1871 certain Pennsylvania refiners, it is not too certain who, brought to Mr. Rockefeller and to his friends a remarkable scheme the gist of which was to bring together secretly a large enough body of refiners and shippers to persuade all the railroads handling oil to give to the companies formed special rebates on its oil, and drawbacks on that of other people. If they could get such rates it was evident that those outside of their combination could not compete with them long and that they would become eventually the only refiners. They could then limit their output to actual demand, and so keep up prices. This done, they could easily persuade the railroads to transport no crude for exportation, so that the foreigners would be forced to buy American refined. They believed that the prices of oil thus exported could easily be advanced fifty per cent. The control of the refining interests would also enable them to fix their own price on crude,

as they would be the only buyers and sellers. The speculative character of the business would be done away with. In short, the scheme they worked out put the entire oil business in their hands.[6]

This brilliantly logical plan broke down before it had a chance of maturing, perhaps because success is seldom granted to those who conceive and canvass a good idea, but is reserved for those who pursue their purpose silently, step by step, taking infinite pains. Where this 'South Improvement Company' failed dismally and ignominiously John D. succeeded only a few years later.

Rockefeller's deep understanding of how a works had to be run and the supreme technical and commercial performance he achieved would not alone have raised him above the standing of a successful business man: what made him the pioneer he became were his insight into the problems of concentration and his method of organizing an industry.

His plan, from the very outset, was to form a nucleus round which his late competitors would rally. Starting with the refiners of his 'own' town, Cleveland, he bought up works which their owners had found difficult to run because they could not keep up the pace of their bigger and better competitors. The compensation which most of them got was pretty adequate, taking into account the works' actual earning capacity in the hands of their present owners, but the value of their refineries in the aggregate was infinitely greater to the purchaser who was not out for *immediate profit*, but was seeking *power* so as to profit to a much greater extent. For a time we see this, if you like, vicious circle: higher throughput of Rockefeller's refineries – greater efficiency – bigger drawbacks from the railways – weaker competitors – higher throughput, and so on and on.

However well Rockefeller played his hand, the freight advantage on the railways would, in the long run, have turned out to be a *tour de force*: for a time it was possible to gloss it over, but the fact remained that *technically the 'unit' of rail transport of crude was the railcar*, and that, if discrimination on the railways were ever to be outlawed, everybody who de-

[6] Ida M. Tarbell, op. cit., Vol. I, pp. 54 *et seq.*

spatched a railcar would be on a level footing, whatever the scale of his freight.

Super Bottleneck: Pipe-lines. At that stage, however, something happened which gave the 'Number One' of the industry the opportunity of leaving all the others far behind; that 'something' was the pipe-line. For, whereas the idea that his total turnover should be taken into account, when a special railway rate for a big shipper was fixed, was based on a *contract*, a pipe-line could, for *technical* reasons against which there was at that time no means of appeal, be used only by firms of considerable size. The paradox in this story is, however, that Rockefeller was far from being the first to realize the importance of this new method of transport; on the contrary, trunk lines were first sponsored by 'Independents', bent on breaking Standard's privileged position in railway transport.

But again it was Rockefeller who reaped where others had sown, not because he had, by some trick or other, managed to steal their crops, but simply because *he* was the one who could make good use of a good idea. As pipe-line economics hinge mainly on constant flow, i.e. on steady and concentrated supply and demand, nobody was to benefit more by the advent of pipe-line transport than the biggest operator. A pipe-line made sense only as part and parcel of an adequate and balanced organization.

Leviathan. For the first time we see the advantage of the complete organization, of what we now call the 'integrated' firm. We are faced with the fact that those prevail who have at their command production and marketing on a sufficiently large scale to take superior transport methods into their service, and who can thus continue to improve their standing at the expense of their competitors. It is very illuminating to study the following statement made in 1939 on the then prevailing set-up by W. S. Farish, one-time President of the Standard Oil Company (New Jersey), as a comment on the position of his spiritual great-grandfather seventy years earlier:

> Integration is the uniting into one business of several of the stages through which a material passes before it reaches the ultimate consumer. The conditions under which integration is desirable are: (1) large volume of business in a single commodity group; (2)

highly specialized production, manufacturing, transportation and distribution techniques; and (3) substantial advantages (at some stages) in large-scale operation. These conditions characterize the petroleum industry, and it follows therefore that the relations between any one of the stages of the industry and the other next to it are peculiarly close. The refiner needs to be assured of his market. The marketer needs to be assured of his supply. Both need a steady flow of products for efficient operation. Neither is interested in other than the one major product and its related group of by-products. Neither can transfer his specialized equipment to the handling of some different product. There is a high degree of mutual interdependence imposed by the facts.[7]

Time and again people have asked themselves how it was possible for one man, starting from scratch, to build up an organization which could twenty-five years later be thus described by the Commissioner of Corporations:

> In the year 1904 the Standard Oil Company and affiliated concerns refined over 84 per cent of the crude oil run through refineries; produced more than 86 per cent of the country's total output of illuminating oil; maintained a similar proportion of the export trade in illuminating oil; transported through pipe lines nearly nine-tenths of the crude oil of the older fields and 98 per cent of the crude of the Mid-Continent, or Kansas-Territory field; secured over 88 per cent of the sales of illuminating oil to retail dealers throughout the country, and obtained in certain large sections as high as 99 per cent of such sales. It also controlled practically similar proportions of the production and marketing of gasoline and lubricating oil.[8]

What began in the Pennsylvania of the seventies developed according to a pattern which we can detect in the oil industry all over the world down to this very day: the ascent of one concern or of a group of concerns which, by centralization of control and by dispersion of interests, attains in due course a paramount position. The two features – centralization and dispersion – are of equal importance. On the former hinges

[7] *Investigation of Concentration of Economic Power*, T.N.E.C. Monograph No. 39-A; 'Review and Criticism on Behalf of Standard Oil Co. (New Jersey) and Sun Oil Co. of Monograph No. 39, with Rejoinder by Monograph Author', Washington, 1941, p. 14.

[8] *Report of the Commissioner of Corporations on the Petroleum Industry*, Part I: 'Position of the Standard Oil Company in the Petroleum Industry', Washington, 1907, pp. xv. *et seq.*

the opportunity of making the weight of large-scale organization felt; on the latter the possibility of taking local difficulties in one's stride. A firm which can average out the result of wisely separated enterprises can never be seriously afflicted by any one mishap, but, on the other hand, it can easily crush the opposition of a competitor, operating from a narrower base, by making a temporary sacrifice which, for the reasons just stated, does not endanger its whole fabric.

Such an explanation of success, mechanical, though in a sense to the point, hardly does justice to a process of much deeper significance. The paramount group in the industry whose power is a thorn in the competitors' sides, and may occasionally be a nuisance to the consumer, plays at the same time an important, nay, an indispensable, role in the industry as a whole.

On Oil Combines

Rockefeller's 'great idea' was conceived with a view to forming a group within the industry sufficiently big to influence decisively the industry's policy and sufficiently broad to survive any earthquake or landslide. *It is obvious that what has been called the 'beneficial' influence of such an outsize organization was nothing more nor less than the natural predilection of the big against violent changes.* He knows that however much he gains on one count, he stands to lose something elsewhere.

Thus the man who heads a big organization will be inclined to take the long view; indeed, he will be forced to do so even at the sacrifice of immediate gains, and all this will certainly make for 'stabilization' or for what has been called 'orderly progress'.

Goliaths and Davids. Throughout the whole history of the oil industry, the central motif is always a leading, sometimes a ruling, group – and clustered around it is a number of smaller competitors. To understand the polarity of *insiders* and *outsiders* is to have a key to most of the mysteries of petroleum. I use the word polarity quite deliberately, since they are not just opponents, they are complementary; the two of them together *are* the industry.

So far I have concentrated on the history of petroleum in the United States because there has been in the U.S.A. such an abundance of published facts and figures covering all phases of the industry from its very beginning. The Americans, money-proud and thus figure-minded as they are, have a singular knack of amassing and digesting material relating to industrial developments, but there is more to it than that. The machinery of congressional enquiries – unwieldy though its derivatives may sometimes appear to the less patient stranger – is one of the features of a vigorously democratic style of life. A 'pressure group' which has to come out into the open is less likely to be nefarious than a gang of conspirators which is allowed to work under cover. In these circumstances, it is not enough to be right, it is almost as essential to know how to state one's case.

If for once I propose to refer to the European rather than to the American history of the co-existence of big and small oil undertakings, this is only because the picture in the United States is somewhat blurred by the legal problems created by the incidence of the Sherman Act. The permanent threat of proceedings under the Federal Anti-Trust laws put a stop to the early development of a fully-fledged organization of oil interests. There is thus more to be learnt from an investigation of the events in countries where the industry was left to work out its own salvation by competition – or by restraint thereof.

Cartels. Whereas, at least up to the first World War, the overall 'control' of the oil industry was in America carried out by one paramount firm, elsewhere the picture was rather of co-operation by a limited number of more or less equal concerns of good standing with a great number of smaller firms, living on the fringe of such organization and still some more existing beyond the pale.

It is superfluous to argue whether associations of companies engaged in the petroleum trade are 'really necessary' – the fact is that they have always existed in every country of Europe. A completely 'free' market was the exception and a short-lived one at that. Untrammelled competition by oil companies in the field took place only during the aftermath of the breakdown of one organization, and represented no more than the period of preparation for a new agreement.

In a survey of 'Cartels: Their Significance for American Business', *The Index* explained that:

> the development of cartels is encouraged if at least four sets of circumstances are present. First, the market must indicate a relatively steady demand. Second, the product must be standardized and easily definable with respect to quality. Third, the industry most liable to foster formation of cartels is that with heavy fixed or overhead costs or large transportation costs, or one which cannot be rapidly adjusted to changing market conditions. Fourth, members of the cartel must have a certain natural inclination for collective agreement and action. If there is a fundamental tendency to overproduction in the industry involved, these factors are apt to become more active. The number of manufacturers of a commodity must not be too large and their economic structure not too different.[9]

What was the object of these 'understandings' between prominent producers, refiners and marketers? Does the traditional charge levelled against such a 'ring', that it is formed with a view to keeping prices up by reducing the quantities available to the consumer, apply in the case of the oil trade? Is such a combine a conspiracy calculated to defraud the population at large, or is it a salutary organization which will safeguard the interests of the community? It can be either, but as it so happens it is more often than not in between the two extremes, a blend of wholesome and harmful ingredients. The members of a cartel are not in business for their health, but that does not mean that there are no healthy features in what they are doing. To quote the same issue of *The Index* again:

> Cartels do not necessarily mean higher selling prices, because business men have learned that efficient production, with a growing market stimulated by low prices, is more profitable than a limited market and high prices. Basic economic lessons of this character are not discarded just because business men form a cartel. In depression periods cartel prices have generally been higher than those of uncontrolled prices, but during boom periods the average of cartel prices has been well below that of the free market.[10]

[9] *The Index* (quarterly bulletin of the New York Trust Company), Vol. XXIV, No. 2, Summer, 1944.

[10] Ibid., p. 32.

Once again it is the tendency to stabilize conditions rather than to drive prices sky-high that we recognize in all these moves.

Not so Restrictive. Whatever the position may be in other industries, one can hardly say that the oil interests have ever inclined to restricting the consumption of their products so as to make them rare, thus confining their use to the well-to-do. But even if spectacular rise in the output of oil is proof of the harmlessness, or at least of the ineffectiveness of restrictive schemes, it remains true that it is difficult to draw a line between a policy of raising prices and one of preventing them from falling below, so to speak, subsistence level. Liefmann's saying that cartels and trusts in general are 'children of distress' is particularly apposite in the case of oil combines, because the petroleum trade has always been a buyers' market. If we eliminate temporary and local shortages from the trend, there was at any given moment more oil readily available than was immediately marketable.

Against this background no restrictive policy, in the full meaning of the word, could have made sense. In view of rapidly expanding markets, the object of all associations of petroleum interests was rather, in the long run, to even out discrepancies of supply and demand than to restrict output.

This tendency towards retrenchment, if not towards restriction, is only the result of one of the industry's unorthodox features: the very fact that the rapid market expansion for petroleum never stopped was the main cause of oil's perennial difficulties. Had a reasonably stable demand prevailed, the supply side would have found its natural level, but the knowledge that to-morrow there would be still more buyers for oil unavoidably caused the entrepreneurs to over-estimate their opportunities.

This state of mind encourages over-investment which can be remedied only by ruthless 'cut-throat' competiton, or by an agreement not to use one's capacity to the full. This sacrifice, considerable as it is in view of the relation of fixed and variable costs, can be justified only if the necessary *quid pro quo* is forthcoming, i.e. if a satisfactory price level can be established.

In our study of the initial stages of Standard Oil's rise to supremacy we saw that its 'monopoly' was first established as a sort of by-product of the ferocious struggle among the several competing railroads; here we have the same phenomenon within the sphere of petroleum itself. The predisposition to make an agreement with your competitor results from the fact that all-out competition leads, in certain industries, in the opposite direction, that is, to a monopoly of one sort or another.

The problems of such 'monopolists', however, diverge widely from the text-book; the operators are not so much concerned with adjusting their output to the level at which the resultant price gives them the greatest possible profit. What matters to them is to achieve a *volume* of trade which gives them the overall advantage of large-scale operations, whereas the actual *price* is to some extent controlled by the lone wolf, the small outsider, the 'marginal' seller. If the big units raised their prices to a level at which their smaller competitors could invade 'their' happy hunting grounds, despite not being so favourably placed as regards rational production and transport and marketing methods, they began to lose ground immediately.

Achilles' Heel. Seeing that their policy of 'price maintenance' is subject to interference by their less potent competitors, the 'Major' group will find it advisable to come to terms with them either by buying them up or by inducing them to join the 'ring' as a full or as an affiliated member.

The characteristics of this eternal tug-of-war going on between the 'Majors' and the 'Minors' are most revealing. The reader will agree that for technical reasons alone the formation of paramount oil concerns was inevitable; their role could not be taken over by a welter of smallish firms. The existence of the 'Independents', on the other hand, provides the ventilation which prevents the powerful firms making it 'too hot' for the public at large. In a way the old grouse of the Majors that buyers tend to take the price from the small firm and to expect the large firm to deliver the goods is not without justification. The Independents, however, are justifiably conscious of their role, and they are wont to denounce what they call the conspiracy of their bigger competitors, but that does not alter the home truth that their livelihood depends more often than not on the very existence of these Majors, that is to say, on the policy the latter cannot help adopting.

First there is the inevitable drawback of big organizations which has always given the smaller men chances to infiltrate: a large concern up to a point takes on the features of a government department, and its executives are liable to become contaminated with what one could call 'the Civil Servant's outlook'. Apart from whatever greater versatility the individual traders or industrialists possess, they can sometimes take advantage of the necessity for nationally organized firms to do business all over a wide area at identical, or at least similar, prices. This inevitably means that the sales in the 'easier' sectors have to carry the burden of costs in more remote regions. The smaller firm can concentrate on remunerative markets and, if the going becomes too heavy, can call a halt without losing face.

Give and Take. Apart from these general factors there remains the very tangible concern the Independent feels for what is sometimes called the 'Combine', whose existence is, from his point of view, so much to be preferred to a free-for-all fight. Nobody can be in a better position than he whose price is protected by the self-denial of others, but whose trade volume is unrestricted! There is no more enthusiastic satellite than the biggest operator outside the ring – but, alas! the more successful he becomes the greater his danger of cutting off the branch upon which he is sitting. For, beyond a certain point, his interference becomes intolerable to the trade powers-that-be and he is faced with the alternative of allowing the 'ring' to disintegrate, and thus losing his former preferential position or of joining the inner circle himself.

Thus, while the position of the biggest 'outsider' is the most desirable, the lot of the smallest 'insider' is the most uncomfortable. He has to put up with most of the snags of cooperation and reaps few of its choicest rewards. In the technical and in the distribution sphere he finds it difficult to keep up with his bigger brothers, yet he is deprived of that natural weapon of the interloper, the under-selling of his competitors. It is therefore not surprising that the big firms very often see fit to cater for this 'lower middle class' by offering them preferential terms of some kind of other.

The Majors *will* make every effort to rope in as many of the smaller firms as possible, because in the case of groups controlling more than two-thirds, or even three-quarters, of an industry terrific strength has as its counterpart extreme vulnerability. If a competitor who commands but 5 per cent of what I am selling can mess up all my markets, then it will be only common sense for me to make considerable sacrifices to keep him from doing so. This is what has been aptly called the nuisance value of the sub-standard competitor. The 'market leaders' will first try to crush a competitor and when, for one reason or another, they fail in this campaign, they will very soon accept him with full honours as a sort of junior partner.

As the only deterrent to such a policy of 'appeasement' there remains always the fear that favourable terms accorded to the convert will be interpreted *'pour encourager les autres'*. The newcomer to the trade is of course the bugbear of vested interests – large *and* small – and if they could be certain that there were no more in the queue, it would pay them to be more generous still towards their existing competitors.

Cartels are Brittle. It has been said that cartel monopolies are preferable to restrictive arrangements, based on customs tariffs, as 'tariffs tend to stick, monopolies tend to break'.[11] The existence of all such arrangements, both of the tacit or of the explicit variety, is indeed not a happy one. The participants are precariously poised on a shaky raft, and not only have to contend with wind and weather, but also suspect that their neighbours wish to cast them to the terrors of the deep. The antagonism of Majors and Independents is not the only problem. There are also the very great difficulties the Majors find in coming to terms among themselves.

As long as all members of that class are more or less satisfied with their own share in the market – the word 'market' being used here in its widest sense, including production,

[11] Lionel Robbins, *Economic Plannning and International Order*, London, 1937, p. 116.

refining, transport, etc. – as long as the 'AS IS' basis is generally accepted, there is every incentive for close co-operation. The problem of dealing with the 'Independents' can be tackled more effectively if there is active co-operation between those who, though they will benefit in the end, have to make some sacrifice first.

The happy days when the spheres are in harmony are, however, interspersed with recurrent periods of a dynamic nature when the demigods elect to contend one with the other and when oil empires are won or lost.

The Ideas of Henri Deterding

Advent of a New Power. The *state* of greatness may be awe-inspiring, but what is really fascinating is the *rise* towards that greatness. To hear how Rockefeller created from scratch the nucleus of what was to be later the world-wide Standard Oil organization stirs our imagination, and to follow the devious path by which a young Dutchman, called Henri Deterding, made himself an international figure and the Royal-Dutch-Shell group a first-class power, is thrilling as well as instructive.

Originally the 'Royal Dutch Company for the Working of Petroleum Wells in the Dutch Indies' was a local company in the Far East, and it was only natural for its managers to sell in the nearest markets, i.e. in India and China, as these areas were at that time huge consumers of American kerosine. In the task of supplying 'oil for the lamps of China' the East Indies fired point-blank, whereas the Standard Oil operated at ultimate range. At that time Standard Oil was a strictly centralized company whose entire industrial activity was concentrated in the States. Its directors, conscious of the virtues of large-scale production, thought that they had no business to produce or refine oil anywhere but at home, and they relied on their striking and staying power to carry them through to success, should a foreign producer choose to challenge their supremacy in any given territory.

'Straight Line'. Deterding soon grasped that the first condition for success against a paramount competitor was not to imitate

his methods but, if possible, to do exactly the opposite. Working on what he later called 'the principle of the straight line', Deterding won the first round by making use of his advantage in the transport field. The fact that Standard – of all people – should have suffered from somebody else's transport supremacy proves once more the old saying that empires are destroyed by the very forces which helped to make them. As late as 1920 Sir Henri made it clear:

> that the advantage of having production not concentrated in only one country, but scattered over the whole world, so that it may be distributed under favourable geographical conditions, has been clearly proven. It hardly needs to be mentioned that the American petroleum companies also realized, although too late, that it was not sufficient to have a large production in their own country. As regards our own group in this respect, its business has been built up primarily on the principle that each market must be supplied with products emanating from the fields which are most favourably situated geographically.[12]

The principle of the 'straight line', however, implies more than transport factors only: it meant to Deterding the vertical control of the whole process from the search for oilfields down to the consumer. In his memoirs he said in so many words that:

> following this straight line in our case simply meant that – by our own efforts and with but little outside help whenever practicable – we, of the Royal Dutch, must set ourselves to bring the oil from our wells to our customer with the minimum of delay... This straight-line policy was, of course, the direct reverse of that of the Standard Oil, whose preference all along was for merchandising on a gigantic scale rather than for the actual production of oil.[13]

Now it is well worth realizing that the fact that Standard kept aloof, up to a point, from production and concentrated on refining and marketing, i.e. that it followed the policy deprecated by Deterding, has always been held up as a particular feat of industrial acumen.

[12] Quoted by Ludwell Denny in *We Fight for Oil*, New York and London, 1928, p. 33.

[13] Sir Henri Deterding, *An International Oilman: As Told to Stanley Naylor*, London, 1934, pp. 50–51.

Another Bottleneck. What makes this difference of opinion so interesting is that *both* principles – Rockefeller's and Deterding's – are sound, and which of the two is applicable depends entirely on the circumstances. Within the framework of the United States oil industry of the seventies it was hopeless to try to 'control' crude production as such, but refining-cum-transport provided the necessary bottleneck. However, at the beginning of the twentieth century we find the position almost reversed. Now, with new fields discovered and developed, not in the heart of an urban civilization as that of the United States, but in far-away and climatically difficult countries, the *'unit' of crude production had become much larger and, for the first time, it was possible to achieve semi-centralized control of essential sectors of crude production.*

Indeed, the advent of Royal Dutch Shell coincides with the introduction of the East Indies, Mexico and Venezuela as important suppliers. In none of these countries, not even in Mexico, did conditions prevail which would permit the small wild-catter to operate in the same way as he did, say, in Pennsylvania or Oklahoma. To obtain a concession great influence and sometimes bribes to the tune of a prince's ransom were required, and to start drilling meant a colossal preliminary outlay on road building and housing, on tankage and pipe-lines, commitments to be undertaken by the strongest groups only. It was the shrewd Deterding and his associates who realized that with the geographical shift of crude production towards the outer perimeter the shape of the industry had changed.

There is still another reason why the Shell people viewed the position in a way very different from the 'classical' approach of Rockefeller's disciples. When John D. appraised the situation he was confronted with a seemingly abundant supply of oil and a still comparatively narrow market; *he thought rightly that the supply of crude would look after itself if he could only control the marketing end.*

When the Dutchman Deterding and his English friends, all born and bred in countries without indigenous crude production, who had lived to see the spectacular rise in consumption in the train of the progress of motorization, when they weighed up their problems, they saw that it was vital to possess the oil, and that *whoever controlled the crude could almost let the disposal of the finished products look after itself.*

The Government – Help or Hindrance? But this is not the whole story. What matters most is the difference between the attitudes of Standard Oil and Royal Dutch Shell towards their respective governments. Again it is not a question of judgment but entirely one of circumstances, each of the two was absolutely right according to its lights and the 'climate' in which it lived and operated.

The traditional hostility of the Federal Administration towards Standard Oil, reciprocated by the latter's calculated aloofness, was due to the response of public opinion to 'trust capitalism', which, it was thought, tended to take rather than to give. However, countries haunted by the fear of oil starvation approved the man who knew how to provide it, and recognized him as the one who gave, never mind what he took for himself in the process.

Until the first World War Standard Oil had little to expect from the Government – for them it was personified merely by the tax-gatherer and the Department of Justice enforcing the Sherman Act – but to Deterding's men and to a few likeminded operators, their Government was a very present help in trouble which afforded diplomatic 'support', financial help, and that priceless moral backing which made the boldest ventures possible.

The 'As Is' Agreement. On these counts the superiority of what we may call the 'Deterding School' is very apparent. It shows a more modern outlook than that of its senior rival. Realization of what really mattered, rather than all the alleged grandiose coups and little tricks of which so much has been made, assisted the Royal Dutch to achieve near-parity with Standard in the international field. The crowning success was marked by the famous though unpublished Achnacarry Agreement between Deterding and Teagle. In this peace pact the *status quo* was made the basis of a far-reaching understanding – it was called the 'As Is' Agreement – and it meant in practice that Shell got away with its stupendous gains of the last twenty years by putting an end to the conflict at the very time when she had little more to gain and a great deal to lose.

Much later, Deterding described the 'burning conviction' of his early days:

> There could be no real business health for anyone of us small Oil companies, unless we co-operated in certain directions, one with another, against the sledge-hammer tactics of our then chief opponent. I urged that some form of mutual agreement between us was all-essential: First, as to placing (when circumstances required it) the production, the transport and the selling of our Oil at definite agreed prices under one specified control; and, secondly, as to serving each market whenever practicable from the nearest source of supply. To my mind, those two points still comprise the first guiding principle of all successful Oil trading. Underlying them was the all-essential factor that we would cut out waste, if only I could bring our competitors to stand in with us. The waste, for instance, which was represented by the duplication of mining and refining plants, pipe-lines and the like, to say nothing of the duplication of transport systems, selling organizations and general administrations.[14]

If Rockefeller's system, however enlightened his methods, was an autocratic one, Deterding's ideas were, in a way, democratic. His group, formed by the co-operation of units of similar standing rather than by one concern engulfing a great number of semi-victims, was the providential promoter of horizontal co-operation.

It is a matter of opinion whether the opportunity for a number of firms to join hands by allocating quotas is not simply a means towards obtaining the advantages of planned production without having to have recourse to a full-scale merger.

Making it illegal for ten firms to co-ordinate their activities may have a result not altogether in keeping with the trust-busters' desires. It sometimes leads to their coalescing altogether, voluntarily or by the weaker parties being driven out or bought up. In a sense, cartels and trade associations are the middle-class versions of trusts.

A New Equilibrium

More Newcomers Still. The principal event which ushered in the modern period of the American oil industry was the forced

[14] Sir Henri Deterding, op. cit., pp. 68 *et seq.*

dissolution of the Standard Oil organization in 1911, which made itself felt in earnest only after 1918. It is interesting to speculate what might have happened if public opinion in the States had been less allergic to the concentration of industrial power in the hands of a few 'czars'. It is certain that the paramountcy of Standard would have been more accentuated than it actually was in the twenties and thirties, but there remains an element of doubt as to whether, legal difficulties quite apart, total control of the industry by one single unit *could* have been maintained. The plain truth is probably that the industry – based, as it was, no longer on lamp oil but on motor spirit and fuel oils – had become far too big to be handled by any one concern. It was inevitable that, in view of the swift and progressive increase of demand and almost equally rapid technical developments, opportunities arose for other enterprising groups to acquire a place in the sun.

It will be realized how conducive a rapidly increasing demand is to the advent of new suppliers – the existing producers will be less likely to put up a stiff fight for 'their' share in the market when their own output is on the upgrade, anyway. It is always easier to capture part of additional demand than to squeeze oneself into a static, or, worse still, a shrinking market.

After 1918 'Standard Oil Interests', as they were then popularly, if somewhat inaccurately, called, were preoccupied internationally by the rising power of redoubtable rivals, and also by a considerable number of good-sized competitors at home. Making certain allowances for the particular legal set-up in the States, we see from then onwards the American scene becoming more and more like the oil industry in other countries as they are described earlier. No longer is one firm in control of the whole industry, it is rather a group of autonomous, if like-minded, concerns which moulds the shape of things to come. *Those who managed to come out on top soon realized how much they had in common.* From this angle the issues on which they could not see eye to eye were of minor importance. Here again things developed along familiar lines: once more it was shown that when a competitor has managed to attain a certain standing, it is no longer politic to fight him. He has to be let in on the ground floor, where he will presently join forces to stave off common foes threatening from without.

A Charmed Circle. I am well aware that there are no hard and fast agreements between the top firms in the American oil industry, and far be it from me to take part in the discussion of this vexed question which has given rise to so much heated controversy in the States, but it is quite obvious that public opinion was guided by a sure instinct when it assumed that *most of the super-firms acted as if there existed an understanding between them.* If this undeniable unison is achieved without clear-cut agreements, it is a still more cogent proof that such co-operation is, to all intents and purposes, inevitable.

The patent position is a case in point. With the advent of thermal cracking in the early twenties, patents became a major issue, and there was considerable friction among the supporters of the several conflicting patent claims. It took the competing parties a number of years to discover that the knot was too complicated to be disentangled, and so they eventually agreed on pooling their patents and exploiting them jointly. Similar developments took place in transport and other spheres of common interest. From that time dates the conception that there is a considerable number of big firms forming a group of 'Majors' who are bound to react in a similar fashion to the problems which confront them and will thus act along similar lines.

Chassez le naturel il revient au galop – whatever we may do, the fundamental factors come to the surface: the oil industry, to exist at all, calls for concerted effort and, however often a co-operative structure may have been disturbed or broken up, it will soon begin to form again.

Control-cum-Competition. Who, then, is on the right track? Those who stress the radically competitive character of the industry, pointing at the almost savage conduct of recurrent price wars, or those who maintain that the industry is under the sway of monopolistic rule, naked and unashamed, or skilfully camouflaged, whichever may be the case?

Certainly there has always been competition, and sometimes of the most vicious kind; it is in these periods that the several units jockey for position. But the relevant fact is that all-out competition could only go to such lengths if it were of a temporary type. Those taking part in the game are wont to make sacrifices far beyond anything that conservative calculation of cost would justify, because *they are not fighting for their share in a free market, but for their quota in the combine which is to be formed eventually*.

On the other hand, those who would have it that the industry is all monopoly must realize that, even if the component parts of the industry seem to work hand in glove, there is always that very strong undercurrent of *potential* competition which tends to qualify the behaviour of a monopoly. Not even the antagonists of the oil powers will accuse them of hampering technical progress or of failing to equip themselves with the most adequate tools for delivering the goods. This is not due to the public spirit of the officers of these firms but is the result of their considering any market against the background of competition, even if transactions are at a given time carried on in an 'orderly manner'.

Hobbes' definition of war as the absence of *real* peace fits the oil world admirably:

> The nature of Warre [he said] consisteth not in actual fighting; but in the disposition thereto, during all the time there is no assurance to the contrary.

In the petroleum industry whose only constant is its changeability, nobody is allowed the luxury of resting on his laurels. However close co-operation among the participants of an 'understanding' may be, every one of them will always keep an eye on the future and the standing of a member within such an organization depends on how well or how badly he would fare, should it break up.

It is therefore certainly unjustifiable to talk of 'oil monopolies'. What really happens is that certain units assume some sort of *leadership* in one or more sectors of the industry. Leadership carries considerable advantages, but they are obtained at a price, as we appreciated during our investigation of the relative position of Majors and Minors. In a way this set-up has so far provided what was required: production and refining and transportation on a large scale which have made for technical efficiency and for decreasing cost. The ensuing 'monopoly' has been tempered by the competition of smaller units and by rivalry – actual or potential – among the major units themselves. Does past history and the present structure of the industry show that, *not being self-adjusting, it can adjust itself*, and that it can always solve its own problems? Can it achieve its own salvation in all circumstances, or do we need a policy *for* the industry?

PART IV

POLICIES FOR THE INDUSTRY

So far I have indicated some of the basic factors in the oil industry, and I have endeavoured to trace its history as a function of these factors. It remains only to investigate how the industry, as we know it, can be made to fit into what is likely to be the general pattern of economic life in the near future. I do not want to preach what we *ought* to do about oil, there each of us is entitled to his own opinion; rather I limit my purpose to the demonstration of what *can* and what *cannot* be done about it.

Patterns for Oil Peace

The literature on international oil affairs – and there is no lack of it – does not really deal with petroleum. Either it consists of more or less accomplished disquisitions on the game's protagonists, based rather on fiction than on facts, or it is straight political argument. This is no one's fault, but it does reflect the fact that, contrary to popular opinion, oil affairs are not the *fons et origo* of what is going on, they are but one of the manifestations of the course things have taken.

Until about 1910 oil was not a front-page political issue, but from the first World War onwards it was so important that it became an immediate concern of most governments. It is true that men like Deterding actually led the field, but such influence as they wielded was in the end due rather to their proving acceptable exponents of their countries' interests than to their control of big commercial groups. It has often been represented that much of the progress of crude production in the Caribbean and the Middle East was due to the activities of individual pioneers. This certainly was the case, but what gave these efforts their particular significance was the support of the Powers. The profit motive is an inducement which must not be underrated, but it remains a subsidiary factor throughout.

Amphibia. The history and, incidentally, the success of the Anglo-Persian Oil Company (later Anglo-Iranian then British Petroleum) is a perfect example of the dual role a big oil concern can, or rather has to, play in our times. It is instructive if only we are able to distinguish between essentials, which the several types of oil companies have in common, and formal set-up, in which they differ. For, while Anglo-Persian was partly state-owned and carried on business in the manner of a commercial corporation, there was, at the same time, more than one privately-owned group which saw fit to act as if it were government controlled. In fact, the difference between the two types of undertaking is much smaller than would at first appear to the layman.

What interests us here in the history of 'Anglo-Persian' are the motives that caused the British Admiralty to take a step as unorthodox as the acquisition of a struggling oil company and its development into a first-class industrial proposition. The whole transaction bears the unmistakable features of Churchillian improvisation and was carried out with that blend of determination and gentleness which, to the outside world, has for a long time been the hall-mark of important moves by the British.

Churchill on Anglo-Persian. The Admiralty, faced with the necessity of relying on liquid fuel for its warships, was worried because its supply was not as safe as that of home-produced coal, and it was Winston Churchill himself, as First Lord of the Admiralty, who described in his statement to the House of Commons, made on July 17th, 1913, the conclusions to which they had come and the policy they had agreed upon:

It is a twofold policy. There is an ultimate policy and there is an interim policy. Our ultimate policy is that the Admiralty should become the independent owner and producer of its own supplies of liquid fuel, first, by building up an oil reserve in this country sufficient to make us safe in war and able to override price fluctuations in peace; secondly, by acquiring the power to deal in crude oils as they come cheaply into the market... This second aspect of our ultimate policy involves the Admiralty being able to refine, retort, or distil crude oil of various kinds, until it reaches the quality required for naval use. This again leads us into having to dispose of the surplus production – another great problem – but I do not myself see any reason why we should shrink if necessary from entering this field of State enterprise. We are already making our own cordite, which is a most complex and difficult operation... and I see no reason, nor do my advisers, why we should shrink from making this further extension of the vast and various businesses of the Admiralty. The third aspect of the ultimate policy is that *we must become the owners, or at any rate the controllers at the source*, of at least a proportion of the supply of natural oil which we require. On all these lines we are advancing rapidly.[15]

Here we have the whole problem in a nutshell: the government of a country which has no crude of its own enters the ranks of oil producers and refiners, not because it believes in state ownership, but because it has no alternative. It cannot afford to rely on the traditional commercial machinery, controlled to a great extent by foreigners, for its most vital supplies. From that time, with one great power's cards on the table, *all* international petroleum developments are found to have lost their private character.

The Dog and its Tail. Foreign activities of American oil firms might not have been the concern of Washington as long as they were, for all intents and purposes, confined to marketing, but in the early twenties the spectre of an oil shortage in the U.S.A. had already raised its ugly head and, however much the opposite may seem to be true, there is little doubt that some sort of co-operation existed between the oil interests and the State Department. To assume that this was not so would

[15] As quoted in E. H. Davenport and Sidney Russell Cooke, *The Oil Trusts and Anglo-American Relations*, London, 1923, pp. 18 *et seq.*

be to underrate the intelligence and the responsibility of either side. Professions to the contrary only go to show that the Americans have now adopted the traditional British claim of coming by the Empire in a state of absent-mindedness.

The problem is not so much one of protecting the property of United States citizens as the safeguarding of the vital interests of the United States. However much the spokesmen of the American petroleum industry may publicize their desire 'to keep the government out of the oil business', they cannot escape history. They cannot deny that the Administration must make sure of potential supplies in case of an emergency, nor that the industry stands in need of what is, by way of a curious understatement, called 'diplomatic support'. Do the oil interests really believe that the Government – that is to say, their fellow-countrymen at large – will see them through in whatever they elect to do abroad, without first acquainting themselves with the layout of the enterprise? Does business really expect the Government to endorse a blank cheque? Do they honestly think foreigners such simpletons as to prefer 'private American enterprise' to state-controlled undertakings when, at the same time, these very oil companies are emphasizing the urgency of vigorous support from Washington? Shorn of all embellishments, which have been designed for tactical purposes only, the relation of state and industry in respect of foreign operations is extremely simple: the oil people are dealing as agents for a principal who has elected to pay them commission on a generous scale. To lay the blame for international unrest at the door of the industry is to confuse the hounds with the huntsmen.

Twilight of the Gods. I am not trying to propitiate the Majors by shifting responsibility to other agencies. In fact, the upshot of these developments is anything but pleasing to them. It is part of a general progression during which they have lost much of their old power, and if things continue in the same direction they may be deprived of their very *raison d'être*.

In the 'heroic' age of petroleum politics, the big oil groups were still considered to be autonomous, even sovereign, powers whoever may have been ultimately responsible for their policies. Like the Elizabethan privateers, they were allowed to carry on their campaigns provided that they kept within the bounds set by power politics.

The wheel has now turned full circle. Deterding's conception that you have to be in with the government if you want the government to work for you has proved almost too successful.

Any government required to shoulder responsibilities will sooner or later claim its rights as well. This being so it is not surprising that the official delegations to the Anglo-American oil conference, held in Washington in the summer of 1944, consisted exclusively of politicians and civil servants while the oil people were confined to talks on a 'technical level'.

By way of consolation to oilmen, smarting under this blow, there is the fact that the expert does not always get the clearest overall picture of an involved problem; it was no less a man than Henry Ford who said he would never employ experts for a new venture, they were too aware of the difficulties. After all, the great men in our industry did not start as producers, as refiners, nor even, if the word is admissible, as oil marketeers. In fact, with apologies to Clemenceau and the generals, oil is too serious an affair to be left to oilmen.

Grand Design. The most vital, if least publicized, function of the Majors since the thirties has been to act as *the great 'eveners' of oil production and distribution.* Their almost complete hold on such critical producing fields as those of Venezuela, Persia and Iraq made it possible for them to open up and shut down production according to market requirements. Only groups with world-wide interests and command of proportionate resources could, for instance, afford to bottle up the Iraq production for so many years. Here we see, though on a different level, the exact replica of the attitude of major firms to the problems of a domestic market as it was depicted in the preceding chapter. The largest operator is, more than anyone else, interested in comparative stability, and he will always be prepared to pay for what he thrives on. It is not difficult to appreciate what would have happened to oil markets in general, if the potential production of the Middle East had been unloaded on world markets in the early thirties.

The same oil which would have been a nuisance in, say, 1935 was a boon to the Allies in 1941. But you have to be a Pharaoh to be able to provide for the seven years of famine during the years of plenty, and only a very big organization could have 'conserved' oil on this scale until it was needed.

There was certainly no agreement negotiated with this explicit purpose, but, if we try to visualize the general layout of forces and interests, we shall see that the comparative peace of the decade before 1939 resulted from a number of nicely balanced *quid pro quos*. I am not trying to unmask a sinister plot of big business against humanity; I believe it is quite the other way round. It is, perhaps, an example of international planning of the highest order and, if there is any criticism to make, it is that there was undue secrecy surrounding the master plan.

Equation of Cost. The basic problem of the era under review is the difference in production methods and production cost between the U.S.A. and the newer oil regions. In the U.S.A. production costs are determined by the high level of its wage bill and the structure of its producing industry, which necessitates the bringing down of a relatively large number of wells.

The average actual cost of producing crude in the Middle East is less than half that in the United States. Had the interests controlling these fields wanted to carry on all-out competition amongst themselves and against United States producers, oil prices outside the U.S.A. would have dropped far below the established 'Gulf of Mexico' standard and U.S.A. export, with the possible exception of specialities, would have come to an end. It is true that the U.S.A. could have settled down behind a tariff barrier but it was apparently found preferable to come to terms with the few groups controlling production in Central America with a view to limiting their oil imports into the States to a certain figure, and to establishing a *modus vivendi* with competitors in foreign markets. This informal agreement with the main producers in the Western hemisphere – outside the U.S.A. – was the relevant factor, although it might never have been concluded, had not the purely U.S.A. interests been able to brandish the tariff weapon.

A policy of basing world market prices on 'Gulf of Mexico' quotations appears to have been generally accepted, and was completely rational after the U.S.S.R. dropped out of the picture, since the Gulf was the main source of supply for independent importers in consuming countries. Taking into account both the low actual production cost in the outlying oilfields and Deterding's famous 'straight line' transport policy, the return obtained for supplies of Middle East and some other crudes must have been highly advantageous, and might have been even more so but for the sacrifices involved in the policy of super-conservation in several fields.

International Aspect of Proration. There can be but little doubt that the American counterparts of this international set-up were 'conservation' and 'proration', as we knew them in the thirties. I do not see in the idea of limiting competitive drilling a sinister attempt of the Majors to kill off the Independents, although it may, in the first instance, have made the latters' life difficult, but no unbiassed observer need deny that conservation suited the Majors' book very well.

The Majors are, as we have seen over and over again, always eager for a certain stability of the market. In this particular instance they want it even more than usual, because they know that sudden outbursts of flush production create circumstances which favour the mushroom growth of smaller refineries near the fields which get their chance whenever and wherever local overproduction brings some crude prices down to a level below the national average. But this is only by the way, the impact of proration upon competitive positions in the domestic market is probably not the only important feature of conservation policy. What mattered most, apart from the aspect of technology, where sound argument seemed to support it, was the fact that only with a certain degree of production control could the United States be fitted into the world-wide structure of the oil industry. Conservation was the missing link which had to be forged.

We have seen that the United States price level – that of a high-cost producer – was maintained with the assistance, the collusion if you like, of low-cost producers who, at the same time acted in accordance with their own interest. They themselves were deeply involved on the American side, and could hardly be expected to allow their foreign interests to interfere materially with their policies in the United States, but it is, on the other hand, pretty certain that the whole structure would have been thrown out of balance had the Americans, say in 1936, seen fit to step up production unduly. The *sine qua non* of a satisfactory price level was, however, a reasonable control of output *everywhere*. The spokesmen of independent producers, who flew into a rage at the very thought of the curtailment of their production, refused to appreciate that, failing an international understanding, their output would have been 'prorated' with the bulk of the Independents forced out of business by crude sold over a long period at, say, 50 cents per barrel. True, supply is inelastic for ordinary price fluctuations, but there exists a certain breaking point at which the whole edifice collapses.

Consequently, the structure of American oil prices was a very peculiar one: those who maintained that, apart from the incidence of proration, prices were allowed to find their own level in the course of competitive transactions were right up to a point, but they overlooked or neglected the fact that *there was an invisible hedge round the American market, formed by the deliberate policy of big foreign producers.*

As long as the U.S.A. cared to export a sizeable part of its production, its domestic market was irrevocably tied to the Gulf price level. Domestic prices could only be substantially higher than export quotations if there was an elaborate system of control over exports and of allocation for the domestic market. The informal system as eventually established was, though by no means foolproof, far more palatable to all concerned.

Had there been no international 'understanding', oil prices would almost certainly have been lower than they were – but is that really a consummation devoutly to be wished? It is always questionable whether low oil prices are at all beneficial in the long run.

The Major Groups had thus managed to establish a world order for oil that worked. Although they were in no small degree inspired by their governments – sometimes, perhaps, prompted, sometimes restrained – they had retained their formal independence, and exceedingly difficult and intricate negotiations had been left to them, each group fending for itself as best it could.

This solution, makeshift though it was, may have been the best that was possible in an era of unrepentant power politics when no truly international organization was allowed to take root.

Washington 1944. Whereas in the preceding period the American and British Governments tended to use the oil groups as tools for their offensive or defensive policies, and on the other hand sanctioned implicitly 'peace' agreements made by the oil interests, it appears that in 1944 they *agreed not to disagree on petroleum* and declared as the firm basis of their policy that every possibility of conflict in long-term and day-to-day developments alike should be eradicated. After 1914 the oil people became glorified agents of the world Powers. Now we move on one stage further – the principals themselves have got together and the agents are functioning 'in an advisory capacity'.

Although the Anglo-American 'Agreement on Petroleum' signed in Washington on August 8th, 1944, was never ratified, certain of its principles embodied in a State Paper could not fail to make history.

The operative clause is, perhaps, paragraph 3 of the Introductory Article, which runs as follows:

> The Governments of the U.S.A. and the U.K. recognise that supplies of petroleum should be derived from the various producing areas of the world with due consideration of such factors as available reserves, sound engineering practices, relevant economic factors, and the interests of producing and consuming countries, and with a view to the full satisfaction of expanding demand.[16]

Some points of procedure covered by the Agreement are important – e.g. that existing rights should not be attacked: that both in the field of supply and of production no

---

[16] *Agreement on Petroleum Between the Government of the United States of America and the Government of the United Kingdom of Great Britain and Northern Ireland*, Washington, August 8th, 1944; H. M. Stationery Office, Cmd. 6555, p. 2.

discriminatory practices should be sanctioned – but what really matters is the decision to get together and to 'suggest the manner in which, over the long term, estimated demand may best be satisfied by production equitably distributed among the various producing countries. . .'[17]

The fact that in 1944 two Governments, whose spheres of interest contained practically all oil reserves outside the U.S.S.R. have concerned themselves explicitly with 'supplies of petroleum available in international trade' and with 'estimates of world demand for petroleum', will be recalled as one of the outstanding events in the history of oil.

It will be realized that once even the very big corporations have to fall back upon national or international authorities their spell is broken; never again will they be treated as if they were in a class of their own and powers in their own right.

Confound Cartels! Surely Mr. Pew, Sun Oil's president, was right when he made a statement before the Petroleum Industry War Council on October 24th, 1944:

> The oil agreement [he said] sets forth objectives which can be achieved only through production control, or control of marketing, or control of prices, or all of these, and I challenge anyone to dispute this assertion. What is that but a cartel?[18]

Cartel, indeed, but what then was the 'understanding' to maintain the world price structure at the level of American stripper well cost? Mr. Pew's chagrin is not really caused by *what* was done in Washington but by *who* did it. What frightens him is not the prospect of an international agreement, it is the spectre of Federal control. Some of the major companies, usually champions of the idea of 'healthy', i.e. limited, competition become restive when they see the Administration adopt their own tenets and President Roosevelt anathematized international cartels.

A few weeks after the stillborn Anglo-American Oil Pact of 1944 was signed, Roosevelt wrote:

> During the past half century the United States has developed a

[17] *Agreement on Petroleum*, pp. 3 *et seq.*

[18] Quoted in *National Petroleum News*, October 25th, 1944.

tradition in opposition to private monopolies. The Sherman and Clayton Acts have become as much a part of the American way of life as the due clause of the Constitution. By protecting the consumer against monopoly these statutes guarantee him the benefits of competition.

And he went on to say:

Unfortunately, a number of foreign countries, particularly in Continental Europe, do not possess such a tradition against cartels. On the contrary, cartels have received encouragement from some of these governments. Especially is this true with respect to Germany. Moreover cartels were utilized by the Nazis as governmental instrumentalities to achieve political ends. The history of the use of I.G. Farben trust by the Nazis reads like a detective story. The defeat of the Nazi armies will have to be followed by the eradication of these weapons of economic warfare. But more than the elimination of the political activities of German cartels will be required. Cartel practices which restrict the free flow of goods in foreign commerce will have to be curbed.[19]

The reply to such a statement can perhaps best be given in the words of an American correspondent of the London weekly, *The Economist*, who wrote at about the same time:

Exchange controls, import quotas, bulk purchases by government, bilateral agreements and cartels (but not commodity agreements) are pretty generally discussed as though they were invented by Hitler. It follows that all nations are expected, upon the defeat of Hitler, to remove these controls with the same enthusiasm which the French Forces of the Interior demonstrated in rising against the German forces occupying Paris. Little consideration is given to the argument that these national controls of foreign trade originated, in part, because of fundamental difficulties in the maintenance of equilibrium in the international balance of payments.[20]

The plain truth is that policies cannot be condemned simply on the grounds that they have been used to attain sinister objectives. Whatever organizations intend to plan for production or trading on a national or international scale will

---

[19] As quoted in *Platt's Oilgram*, September 8th, 1944.

[20] *The Economist*, October 21st, 1944, p. 570.

have to make use of a technique similar to that evolved by cartels.

Once again the most lucid interpretation of the problem is to be found in the columns of *The Economist*:

> It is no more possible in the international sphere than in the domestic to argue in terms of black and white. To condemn cartels is one thing, but to condemn all forms of economic organization that include the exercise of purposive direction over production is an entirely different thing. In the international sphere, as in the domestic, it is necessary to distinguish and to define.[21]

London, 1945. The trend of events during the year that passed between the signature of the 1944 Oil Agreement in Washington and that of its watered-down second edition in London does not really cut across what I have said. The objections to the original text were directed against its ambiguity which disturbed those who believed, rightly or wrongly, that it was but the façade of a far-reaching division of supplies and markets. Those, however, who would have accepted more concrete commitments, had they only been properly set out, had the ground cut from under their feet by the protests voiced in the United States by those who refused to enter into *any* working agreement likely to prejudice their freedom of action.

Thus the agreement signed on September 24th, 1945, differs from the one of August 8th, 1944, in that it has apparently no teeth left. So as to make it acceptable to Congress and to dispel the misgivings of American producers, all clauses which could be construed as affecting domestic economic factors in the States have been removed and what remains looks in the first instance as if it were nothing but a string of pious platitudes; indeed, a wit described the Agreement as 'a Japanese kimono which covers everything and touches nothing'.

True, lip-service is paid to the desirability of 'petroleum being accessible in international trade on a competitive. . . basis', but elsewhere there is another reference to the need for 'efficient and orderly development of the international petroleum trade'. 'Orderly' is the operative word: no sooner was

[21] *The Economist*, December 2nd, 1944, p. 725.

the Agreement signed than Mr. Shinwell, the British Minister responsible for petroleum, pointed out that it 'will introduce some measure of order into the industry throughout the world, which will be to the advantage of all of us.' None of the participants in the game is likely to forget the lesson of the all-round benefits they, and not they alone, had derived during the period of 'orderly' development in the thirties.

There is an inescapable logic of events which can be camouflaged for the sake of pacifying certain powerful interests, but which will prevail none the less.

Competition and Control

Sooner or later the student must come up against this crucial problem: which form of organization will be the most beneficial to the oil industry from the long- and short-term point of view? Here this general question can be particularized in two fundamental questions:

(1) Is it more likely that supply and demand can best be brought in line by free competition or by its restraint?
(2) If some 'planning' is deemed necessary, should it be left to the industry itself to devise ways and means, or is it definitely a public concern?

Was Competition Ever 'Free'? Unless my reading of the oil industry's structure and history is altogether wrong, there is no question that there has been, always and everywhere, an overwhelming tendency towards concentration, integration, and cartelization in the petroleum industry. This goes far deeper than Adam Smith's taunt that:

> people in the same trade seldom meet together, even for merriment and diversion, but the conversation ends in a conspiracy against the public or is some contrivance to raise prices.

It is, as I hope I have proved, due to the very fact that all-out competition, where it is allowed to prevail in the oil industry, leads either straight to general bankruptcy or to the monopoly of a survivor.

There can be little doubt that competition – spontaneous or even 'enforced' – results in the formation of a 'hard core' whose monopolistic tendencies are more or less tempered by actual competition from outsiders and by potential competition among the leading members themselves.

I should like to refer those who are not prepared to accept historical evidence as sufficient proof for the validity of an economic doctrine to the current beliefs on the most adequate method of exploiting exhaustible or semi-exhaustible raw materials – what has for a long time been the basis of enlightened forestry has now become the recognized practice in working other natural resources. In one of the monographs, commissioned by the Temporary National Economic Committee (T.N.E.C.), dealing with various aspects of 'Concentration of Economic Power' in the U.S.A., one can find the following passage:

> Competition contributes to efficiency in manufacturing and in distribution; it causes inefficiency in the utilization of natural resources. Competition in the production of timber, bituminous coal, and petroleum hinders the the application of improved technology and encourages the employment of wasteful methods of exploitation. It may provide the consumer with a large supply at a low price for the time being, but it does so at the expense of future generations. Competition is not conducive to conservation. Where competition does contribute to efficiency, the gain is offset, in part, by the wastes which it entails.[22]

Natural Monopoly. The petroleum industry has been described as an:

> activity that would be expected, from a purely physical standpoint, to function with maximum efficiency as a natural monopoly.[23]

Should we be prepared to accept, if only for argument's sake, the statement that integration, concentration, and co-operation are endemic features of the oil industry, we may perhaps also agree on the conception that *the organizing forces within the industry are not 'restraining' competition which would otherwise be 'free', but rather developing a peculiar blend of monopolistic and*

[22] *Investigation of Concentration of Economic Power*, T.N.E.C. Monograph No. 21. Clair Wilcox, 'Competition and Monopoly in American Industry', p. 14.

[23] Joseph E. Pogue, *The Economics of Petroleum*, New York, London, 1921, p. 3.

competitive tendencies whose interplay has shaped the industry from its beginnings to this very day.

This still leaves the second question: who should do the planning? Here again there is no patent solution, no panacea for the 'sea of troubles' besetting the oil industry.

The answer to the question whether or not the oil industry in a given country is satisfactorily organized depends entirely on the principles guiding the policies of that country. There the behaviour of oilmen, as in the international field, cannot be judged without taking into account the state of affairs prevailing generally at the time. If the very principles of communal life are under review, any change on that level may outlaw practices which were formerly quite acceptable, and any change of premises, as it were, may provide opportunities for action which hitherto did not exist.

It is rather likely, for example, that a government monopoly of oil production or refining might turn out to be a doubtful venture in a country where the *general* constitution of trade and industry conforms to the pattern of competitive activities. To single out petroleum as a 'key' industry for special treatment, to sever its connection with the bloodstream of 'capitalist' business, is asking for trouble. This would, in a way, mean making the worst of both worlds. Such an oil monopoly would enjoy neither the advantages of fully-fledged planning, nor the impetus provided by the rivalry of competing units.

It is beyond doubt that the oil industry owes much to private enterprise, to the pioneering spirit of those in search of fresh fields and markets new.

The merits of individual oil enterprise are not altogether confined to the producing side. The pieces which the oil industry has to render in countries without total planning are rather too intricate to be played by a brass band. The world would be poorer and a duller place if the keen fire of rivalry was to be smothered by government autocracy, however enlightened.

Limitations of the Independents. On the other hand, if what I have said about the structure and the history of the oil industry shows anything, it is that competition has in every respect and all along been subject to a high degree of voluntary or compul-

sory control. Those who have, sometimes with every justification, preached the intrinsic virtues of a firm leadership for the
industry, will find it difficult to deny that petroleum lends
itself to organization round a few focal points. There is a
tendency among the spokesmen of 'independent' interests to
shut their eyes to the almost overwhelming tendency of the
industry towards concentration. Whilst they are eager to give
a vivid account of the shortcomings of the present set-up, they
are never equally articulate when it comes to saying how the
Independents could possibly run the industry without themselves becoming 'Majors'. Nevertheless, their argument is in
keeping with their basic principle that the 'unit' of the industry must be kept comparatively small. The spokesmen of the
Majors are up against a far more formidable difficulty. Their
stock-in-trade is the need for big units, for integration, for
large-scale operations, and long-term policies. From such a
platform the fight against planning by other agencies – national or international – is a much harder one.

*Dual Role of the Majors. The justification for the existence of the
Majors is that they play an indispensable role. It does not consist in their
being just one of the competitors, but in their being so big that at times
they cannot help thinking and acting as leaders in the true sense of the
word.* I have told the story of the host of associations of oilmen,
operating on an equal level, who could not find a form of
organization that would outlast the first storm, because the
interests of each one of them were identical to such an extent
as to be, in the end, incompatible. The very big man, however,
can contrive to live side by side with smaller units. He can
keep them in line up to a point, and can afford to make
valuable concessions designed to secure their agreement. By
thus forming the nucleus of an organic structure the Majors
are, in a way, acting as authorities who, however much they
may be pursuing their own ends, function as trustees of the
public. Much as the Majors' performance of these duties is
appreciated, it remains true that they could, given certain
circumstances, be performed as well, or better, or at least
more equitably, by agencies responsible, not to their shareholders, but to the public itself. If in an industry there is a
need for a certain degree of regimentation, for what the

Americans call a 'Code', it will be more readily accepted if it is devised and enforced by an authority with no axe to grind.

The Majors have fully realized that a certain regulation or stabilization – if that term should be more acceptable – of crude supply could not be achieved without governmental influence. The contention that proration was nothing more than a scheme of the big groups to re-establish monopoly is certainly an over-simplification. Does not the opposite view, that the role of 'evener' had to be taken over by some authority other than the Majors, offer much more interesting possibilities?

Economics of Proration. Proration, as it has developed in the United States, is the foremost example of what can be achieved. Agreements between producers have always been thought to be necessary in some form or other because of the poor price-elasticity of crude. The record of these understandings, the participants themselves seeing to it that they are not in the long run viable, proves them to be, in Hobbes' words, 'nasty, brutish and short where there is no power to overawe them all'. As early as 1924, when President Coolidge took the first timid steps towards conservation by constituting the Federal Oil Conservation Board, he thought it necessary to refer to the peculiar relationship of industry and government. He first paid lip-service to freedom of enterprise by saying that:

> the oil industry itself might be permitted to determine its own future;

but then he went on:

> The future might be left to the simple working of the law of supply and demand, but for the patent fact that the oil industry's welfare is so intimately linked with the industrial prosperity and safety of the whole people, that Government and business men can well join forces to work out this problem of practical conservation.

Public interest, however, was not strong enough to force even a limited measure of control down the throat of a suspicious industry, and it was only the repeated and unmanageable glut of the late twenties and the early thirties which left it with no alternative other than the calling in of the administrative powers of State Governors.

Proration is a first-rate example of the marrying of communal and private interest. The general outline is worked out by an authority, but within this framework the actual job was done by private enterprise, and the efforts, the skill, the resourcefulness – and the resources – of the competitors still determine their respective standing.

Imperfect Competition. It does not make sense to object to 'state interference' or to regulations, imposed by agencies outside the oil industry, in the name of 'orthodox' free competition. My narrative and description of the structure of the petroleum industry shows that there has prevailed, always and everywhere, *a state of monopoly, qualified by competition*, or, if you prefer it the other way round, *a state of competition, qualified by monopolistic control.*

Although the oil industry may be one of the foremost examples of such 'imperfect competition', its problems are, in general, those of contemporary industry. Oil may, however, differ from some other industries in so far as it nearly always *was* organized in such a way as to have some sort of 'quantitative control' applied and allowance for the 'rigidity' factor was made by market leaders, big enough to be 'rigid' without disintegrating under a sudden or unexpected impact. *The problem is therefore not for the authorities to enforce a new system but, if and when necessary, to take over some of the functions so far fulfilled, more or less successfully, by the major companies. Whoever is going to exercise control over the petroleum industry will be in the happy position of carrying on a great and deep-rooted tradition.*

It may be asked, not without justification, Why change? The answer will be that, though in a *laissez faire* world the particular brand of competition-cum-control developed by the petroleum industry would appear to be entirely adequate, it is an altogether different proposition to fit it into the current trend of economic life. The scene was set for such a change some time ago: the rising influence of governments in international oil deals, export and import tariffs and regulations for the location of industry.

Full realization of the new relationship of government and

industry in general is bound to lead to further progress in the same direction. It is hardly possible to appreciate conditions in any one industry without taking into account the prevailing state of affairs in the others. The salient fact is that governments have no alternative to concerning themselves directly with industrial developments, once they have made themselves immediately responsible for their peoples' 'pursuit of happiness'.

Limits to Laisser Faire. I have never had the slightest doubt that the thesis that a maximum of goods would be made available with the least cost if economic factors were left to find their own level without any interference from other agencies, was irrefutable as far as it went. The problem, however, is, just how far it goes, or perhaps how far we can afford to let it go. Such conception presupposes a complete interchangeability of materials and men and it cannot function without a continuous and sometimes ruthless elimination of certain means of production in favour of more efficient or effective competitors.

None of these conditions ever obtained completely, but in the nineteenth century people, and capital for that matter, could migrate comparatively easily to lands where work or investment yielded greater benefit, whereas in our century both have become immobilized by the rising tide of nationalism. Another point is more momentous still: the system of 'free' economy works only if all elements of production can be treated alike – if I find that a piece of plant does not any longer fulfil its purpose, I scrap it; and if the workmen are no longer worth their pay, I sack them. Within the framework of an expanding economy this may be of no great moment, involving nothing but a swift change-over from one job to another; in different circumstances, however, and as a mass occurrence, the problem becomes insoluble, as far as the individual wage-earner is concerned. His plight makes us realize that the worker is not just an 'expendable' tool, and that his welfare cannot ultimately be subordinated to what used to be called 'purely economic considerations'.

Once it has thus been found impossible to apply such considerations to the human element in industry, selective elimination which entails the survival of the fittest ceases to operate. Once we elect to consider *one* element in the process of production as taboo and decide to 'fix' it, we shall in due course have to fix or, at least, control *all* the other elements as well.

No Feudalism. If any control exercised in oil affairs, undertaken in the process of over-all planning of investment and production, is to embody the cardinal advantages of the traditional set-up it will have to be restricted to the minimum and so give individual enterprise as much leeway as possible.

The acid test of any arrangement of this kind will always be its success in maintaining the incentive for efficiency and progressive thinking despite the unavoidable integration into a general plan. A modern feudalism with bounties bestowed upon pressure groups and patronage meted out to gangs of political camp-followers has no chance of survival, but it should not be beyond our capacity to devise and maintain a system which ensures the necessary minimum of conformity with a maximum of liberty. In other words, it is neither the state-run monopoly nor state-controlled public corporation which is likely to give the best results in our highly diversified industry, but rather a system which creates certain conditions under which the men on the job can compete by making use of their wits no less than their skill.

Combined Operations. Having, as it were, outrun my lines of communications, I cannot go any deeper into the problems of industry and public control, but there are many more instances of both parties reaping the benefit of genuine co-operation. Such a policy will be successful only if, instead of imposing a ready-made regime upon a trade, infinite pains are taken to adapt the measures, designed to serve the public interest, to the traditional features of the industry. In the realm of petroleum that will mean accepting the big units as working entities – *it is no good trying to grow wheat in the backyard.* The prevalence of vertical integration and of horizontal associations will also have to be recognized. This book has shown that such cartels as existed tended 'to regulate, but not to abolish competition'.

Governments, for domestic or international reasons, have to insist on certain basic regulations being enforced. There is, however, no reason why outside that sphere there should be no incentive for an enterprising producer, for a genius of organization, or even for a wizard in salesmanship. I have never heard that because the bee-keeper provides a sort of prefabricated beehive his swarm is any less efficient than it used to be when housed in an old tree.

There is one crucial test, however, for all schemes to regulate industry: that is their attitude towards the newcomer. Any order, old or new, which makes an industry a closed shop is essentially unsound. Ossification and eventual atrophy are inevitable when industrialists and traders no longer feel the wholesome sensation of newcomers treading on their toes.

Conclusion. To sum up my argument:

As there is always either too much or too little oil, the industry, not being self-adjusting, has an inherent tendency to extreme crises; this fact has called forth the ingenuity of planners within the trade. As no individual unit can evolve a rational production policy on its own, some sort of communal organization is almost inevitable. Paradox though it may appear, oil, competitive *par excellence*, is usually controlled by some 'leading interests'. The Major companies have in the past played a vital part, with the Independents as an indispensable corrective, but now their role is being taken over step by step by other agencies.

In the international field governments are severally and jointly fulfilling the function of organizing the industry. In the domestic sphere the authorities will be compelled to act as eveners and stabilizers, thus taking a leaf out of the great oilmen's book.

Such procedure will be a success only if control, while ensuring adherence to an overall plan, leaves intact individual enterprise of the industry's component parts. Whereas decisions of a *strategic* kind cannot be evolved but on the highest level, *tactical* decisions are best left to the industry itself.

2 How Much Competition Is There?

Competition, whether there is too much or too little, has provided a continuing and often emotional debate for analysts, economists, oil company spokesmen and bystanders throughout oil industry history. Here is a typically cool assessment of the state of competition in the US oil industry in 1946. It is one of a series of seven articles Paul Frankel wrote for the Petroleum Times, *at that time the leading magazine of the industry in the UK, during May–August 1946 after his first post-war visit to USA. This article provides a practical commentary on what Frankel treated in a more theoretical manner in* Essentials of Petroleum *which was published in the same year.*

Is the oil market still 'monopoly-ridden', as Mr. Churchill dubbed it more than thirty years ago, or is it by any chance the domain of perfect competition? Is the industry under the spell of the 'Combine,' of a basilisk against which the small independents fight gallantly if not altogether successfully, or are the big companies standard-bearers of public welfare after all?

The more one knows of oil in general and of the American petroleum industry in particular the less will one be inclined to assess it in headlines and slogans more appropriate to political affrays than to discussions in terms of economics.

The truth is that the 'major' companies are less powerful and the small ones less independent than one would have thought. There is certainly nothing approaching a monopoly in the traditional sense but, on the other hand, competition is not quite as hot as it is sometimes represented to be.

The actual life of the industry proceeds to an intricate pattern made up of monopolistic and competitive, of restrictionist and expansionist elements, and it is this very variety which makes its study a fascinating adventure.

Petroleum Times, June 22 1946.

Majors v. Independents?

The present state of affairs can only be understood if we remember the peculiar history of the industry, which was for more than one generation dominated by an overwhelmingly powerful Standard Oil Co. whose share in refinery throughput amounted to as much as 85 per cent. in 1904. Competition then consisted of the struggle of small units for their bare existence and in those days an 'Independent' was any operator whose business was not owned or controlled by Rockefeller's Trust. The dissolution of the old Standard Oil Co. which did not become properly effective until after the last war and, no less, the stupendous growth of the industry during the advent of the 'gasoline age' created an open field and made it possible for a considerable number of newcomers to occupy seats in the front row. Among these were companies like Texas, Gulf, Cities Service and Phillips, indeed recently the eight 'Standard' companies refined less than 40 per cent. of the U.S.A. total, as against the 45 per cent. of the *nouveaux riches* companies which had graduated into the charmed circle of the 'big twenty,' leaving 15 per cent. to about 230 smallish refiners, still called 'Independents.' The comparative shares are, incidentally, quite different at the producing end of the industry where the 20 biggest companies control only about half the domestic crude oil output.

It has often been said that it was arbitrary to put the twenty biggest firms into a special category of its own, to lump all the others together and to assume that the majors formed a block opposed to and opposed by 'the rest.'

When I met Fayette B. Dow, in Washington, general counsel of the National Petroleum Association, an organisation which, dating from the early days of anti-Standard campaigns, now represents interests in the older oil regions, he told me about the mutual relations of the strata of the industry.

He admitted that there was 'competition among the majors which was genuinely intense' but he added that, from the angle of the independents 'they are nevertheless classed as a group.'

Other unofficial spokesmen of independent interests have

gone much further and one of them published a book before the war under the title *Return of Monopoly*, in which he endeavoured to prove that there was some sort of conspiracy of big business to the detriment of independents and consumers alike. The real facts however are less dramatic but much more significant.

Two Types of Competition

If the top-ranking companies do act in unison although there are undoubtedly no contractual obligations to that effect, this serves as a conclusive proof that they have enough in common to make them think along similar lines as a matter of course. This relationship among the majors has not always prevailed, indeed, it is characteristic only of the *static* periods in the history of our industry, of times when each of the competitors was inclined to accept his comparative standing among the others. There were, however, in the past, and there may be in the future, some *dynamic* spells during which there took place all-out fights for 'living space' in the oilfields and for a 'place in the sun' in respect of markets.

The confusing fact is that both species, the conservative and the aggressive, are called by the same name of competition. They are in many ways different yet they have some elements in common – in fact they represent different stages of a continuous process.

The discovery of a new field, the ambition and the skill of one man, or even strictly political events, disturb an existing equilibrium and the subsequent fight for standing temporarily creates a war of all against all, in the course of which factors separating the competitors weigh more heavily than the interests they have in common. It is this catch-as-catch-can affair which is commonly believed to be the genuine brand of competition.

There comes, however, sooner or later a moment when the weaker of the contestants have been knocked out and when the law of diminishing returns begins to tell on the finalists – it is then that a new balance of power begins to emerge and that, in fact, a different type of competition comes to the fore.

This 'more adult and settled phase,' to quote once more

S. A. Swensrud's felicitous formula, is perhaps best described as one of *potential*, as distinct from virulent, competition. We know it particularly well because, on the international scale, it has been prevalent ever since the late twenties until now, or at least until the incident of the American-Arabian pipeline project of 1944.

In the American domestic orbit a similar development took place – for the last fifteen years there has not been any material change in the relative positions of the teams, indeed, a future historian will summarize that period as the 'As Is Age.' This, however, does not mean that all was plain sailing. In our industry, overshadowed as it is by the Great Unknown – the advent and the fading out of oilfields – nobody is ever allowed to rest on his laurels. However cooperative the companies are in their desire to further interests they share with the others, they devise their long-term policies in such a way as not to be outdone in any respect. True, the battles they fight with each other are minor engagements not designed to make much difference either way, but they provide an excellent training ground on all important levels – exploration, research and marketing. One can say that even *during* the static periods those companies will fare best which have least to fear from a recourse to dynamic methods.

The marketing side of the industry gives a particularly clear picture of the actual functioning of potential competition: the fallacy that competition consisted of selling cheaper than the next fellow has long been exposed, and there is something in the argument that the main products are sold in a given area at almost uniform prices because there is competition and not because there is none. The fact that no one can afford to sell much dearer than the others makes for a common price level. American experience over the last twenty years has proved that the price indices of petroleum products kept well below those of other manufactured goods, which shows that prices have been levelled down rather than levelled up.

Majors *cum* Independents

It is, however, impossible to look upon the market developments as being determined only by the policy of the biggest units. They form the nucleus and hard core of the industry but there is grouped around them a very great number of smaller operators whose influence goes in many respects far beyond their weight in actual tonnage. Having described the peculiar blend of co-operation and competition, both of them genuine, of which inter-major relations consist, I now turn to describing the co-existence of Majors and Independents.

Here again I have learned to disbelieve the popular conception that there was an inevitable and complete antagonism between the two levels of the industry – a conception which dates back to the old Standard Oil days of rough and ready tactics. To-day things are indeed, more complicated. First of all the majors fully accept, nay appreciate, the existence of independents. The latter not only perform an invaluable service to the whole industry by their wildcatting activities on the producing side – 'the credit for the discovery of America's oil fields goes to the small exploratory enterprise rather than the large organisation' Mr. Farish once said – also they are, for the majors, some sort of collapsible extension of the main structure. While they thus provide the outlet for occasional surpluses of one major or another, and the no-man's land, as it were, whose existence helps to iron out peaks, their main importance lies in their standing between the big units and direct public control of the industry.

It was one of the theses of Karl Marx that concentration of industry would inevitably lead towards control by the community and it was not for nothing that, the other day, the Chancellor of the Exchequer called Sir Andrew Duncan, Conservative Member of Parliament for the City of London, 'a great Socialist executive' on account of his success in organising and unifying the electricity and the steel industries. It is obvious that the smaller the field becomes, the easier it is to take the ultimate step and to consider the industry as a unit simply by taking a leaf out of the big companies' book.

Far be it from me, however, to dismiss the independents as stooges for the majors. I have been an independent myself all through my business career and I believe, with all due respect for the achievements of the big groups, that we can rightly claim we perform a public service. The independents, continuously on the look-out for an opportunity of infiltrating into the markets, are wont to detect and are never slow to make use of any unduly high margins which may result from major company policies.

They are thus a most effective check and, however much they are animated by their own business interests, they provide, at the same time, the ventilation which prevents the bigger companies from making it too hot for the public.

Yet there is, in America, a great deal of co-operation between majors and minors. When I saw Fayette B. Dow, he was just making a survey of the status of the independent refiner, and he said that 'the record indicates that the competition, although tough, has been progressively cleaner. There is no charge of malice. No one has written to me suggesting that the executives of the major oil companies are sitting up nights planning evil ways of putting the independents out of business.' The independents, on the other hand, know full well that they are operating within a framework established by the majors, they fill in the gaps and on the whole they are not faring too badly: they sometimes buy from and sell to the Majors, they benefit in a way from the technical progress made by the big interests and still more from the sort of clumsiness inherent in millipedes, which gives an advantage to fast moving ants.

'Loyal Opposition'

I fully appreciate that the workaday life of the independents is not as idyllic as all that and that there is much struggle and hardship, but such difficulties as there are, are confined rather to newcomers who, human nature being what it is, are unwelcome to majors and established independents alike. Once, however, an indepedent has weathered the initial storm he is willy-nilly accepted as a junior member of the fraternity and joins in with the others in being a kind of 'loyal opposition' to the majors.

I am deliberately using language of a kind normally covering constitutional matters because my stay in the States has confirmed my conception of the majors as, owing to their very size, inevitably acting as 'Authorities' of the industry – to quote F. B. Dow once more, 'the integrated companies have become institutions.'

In force to-day we see a system in which the big companies form a concert – and any student of history knows that under the 'Concert of Europe' co-operation of the great powers did not impair but in fact safeguarded freedom and prosperity of the smaller nations.

The specific role of the American major oil companies in respect of the independents has best been described by one of the men I saw – he is unique inasmuch as he shuns rather than seeks publicity, and did not want to be named – who emphasised that what the independent producer needed more than anything else was the knowledge that he would always have (at a price) a purchaser for his production. My informant said that it was as 'Common Purchasers' that the majors had become the industry's backbone.

Looking at the picture as a whole, we cannot fail to agree that there is competition in the American oil industry and that there was never a sign of the concomitants of monopoly, of the tendency to halt technical progress, or overcharge the public grossly. This brand of competition is, however, qualified by the co-operation among the competitors which has been called forth by the need for a greater stability than could be attained by more 'orthodox' methods.

Collective Common Sense

What they have in America is an involved system resulting in a fine balance of the several interests, a balance which is almost automatic due to its counter-checks and built-in safety valves. To the stranger it looks as unwieldy and illogical as the English educational system, but like the latter it has all the resilience of natural growth.

When, in 'A Midsummer Night's Dream,' Bottom the Weaver proposes playing the lion, he promises to roar 'as gently as any sucking dove.' He knows that a stage lion has to roar, but must not overdo it, lest the ladies be frightened out of their wits.

The industry has thrived on the lion competition having been taught to roar gently. This state of affairs is, as one of the big crude producers told me, simply due to 'the collective business common sense of the several competitors.' This brilliant formula should help us to understand a good deal of what is going on in America – and elsewhere.

3   Oil: Perils and Possibilities

Paul Frankel has been an assiduous journalist, as well as an analyst, of the oil industry throughout his career. Journalism has for him been part of the process of educating people about the nature of the oil industry and its participants. During the war he did much work for the BBC External Services and after the war was for many years a frequent speaker on both the Home Service (as it was then called) and the Overseas Service. This piece, broadcast in June 1956 a few weeks after the nationalization of the Suez Canal and before the Anglo-French-Israeli invasion of Egypt in October, gathers under one heading several of the themes that have provided continuing fascination for Frankel – supply/demand, investment, security, profitability. It was written when nuclear power was seen as a promising alternative to oil, but Frankel's optimism for the oil industry was unimpaired.

On previous occasions when I talked about oil on the wireless it was usually about some political crisis, such as the outbreak of the Iranian trouble or its solution by a compromise of sorts. What I have to say tonight has nothing to do with cloak and dagger stories, yet you will find a good deal of drama in it: oil is in the thick of the new industrial revolution which appears to be going on just now.

Hitherto petroleum products have been used mainly as fuel for vehicles – petrol for motor cars and aircraft, diesel oil for lorries and fuel oil for ships and locomotives. The increase in the use of road vehicles and of aircraft over the last fifty years has been stupendous and uninterrupted, yet it took place over a period of fifty years. It took so long because oil consumption for road vehicles and aircraft grew *with* the expanding economy and not at the expense of another supplier. Therefore one had to wait for roads to be built, for cheap and reasonably

BBC Home Service, 28 June 1956.

breakdown-proof vehicles to be designed, and for long-wearing rubber tyres to be developed.

Now however the position is different: with industrial activities growing at a pace fast and furious, the traditional sources of energy – coal and hydroelectricity – are falling behind requirements. Now, almost overnight, oil is called upon to take over part of a vast *existing* market. In the years to come a substantial part of our electricity, of our town gas, and of the steam raised in our factories will come from oil. True, all transport other than that by rail has been based on oil already, but from now on our way of life and our standard of living will be still more immediately dependent on oil.

The question arises: will supply be equal to demand?

You will at one time or another have heard of crude oil in the ground being a wasting asset, an irreplaceable gift of nature which we should use sparingly – indeed not so long ago the world appeared to have oil reserves for not more than fifteen or twenty years at the rate it was used then, whereas coal reserves were expected to last for some hundred years. This traditional picture now looks to us all upside-down: the coal reserves cannot be put to use adequately because the miners are not there to get it, while on the other hand scientific progress in locating oil and in drilling for it has (with a great deal of good luck) resulted in vast resources of oil being available for a very long time to come. These new areas of prodigious production are Venezuela and, more particularly, the Middle East. Their development came just in time to match the growing demand, and helped to remedy what might otherwise have been a very ugly situation.

But some of the 'new' oil regions are in under-developed areas where, in the absence of urban and industrial life, there is little oil consumption and less political stability; on the other hand Europe, a great consumer of oil, has only a small production of its own, and even the United States has now become an importer. Oil is in fact a displaced commodity.

This has made people anxious about relying too heavily on Middle East supplies – only recently it has been said that there would be millions of unemployed in Europe if the flow of oil from the Middle East were to stop; happily the governments of the Middle East countries know that their own income would stop at the same time. Such knowledge usually 'concentrates one's mind'. A major political conflict may result in a stoppage, but then the West would most likely have its own remedies. In peacetime the supplier needs the customer at least as much as the user depends on the producer. One cannot discount some of these threats altogether – we have seen nations cutting off their nose to spite their face, but rarely will they go so far as to cut their own throat. For that is what a stoppage of oil exports would mean to the Middle East countries. Indeed, it would be in the interest of Middle East countries to establish an atmosphere in which the consumer countries would feel safe. Any doubt about their free access to Middle East oil on fair terms will encourage the investment in exploration and drilling elsewhere, even at high cost which normally would make such oil uneconomic. Once such oil is found in a consumer country that quantity is finally lost to the Middle East as an exporter because countries *will* make the best use of their own production once they have it, and will restrict imports.

If we can therefore rely in peacetime on Middle East oil supplies, the position would seem to be reasonably balanced. Perhaps it is even more than that. In oil there is never a dull moment, and just when it looks as if it was coming into its own there appears on the horizon a new contender.

During this century everything had conspired in favour of the oil industry; whatever happened in peace and in war has widened and deepened its scope; the phenomenal development of road and air transport, with a host of subsidiary events such as the use of bitumen for road-building, of solvents for paints and dry-cleaning, and then again the motorization of agriculture, with the use of oil for making most of the new chemicals as the crowning glory — indeed to use an American term: oil was the 'Last Frontier' of our industrial life, the 'real McCoy' of progress and expansion.

The new contender, as you will have guessed, is 'Atomic Energy'. For the first time a major new development does not involve oil, indeed atomic energy may one day compete with or even supplant oil in some of its main uses. We know on good authority that atomic energy will not be a significant source of energy for the next ten years or so, but this is obviously not all that matters in this story. The fact is that by developing the technique of producing nuclear power the lid has been taken off a very big pot and what we are witnessing today may be just a beginning. Much of this might have been another fifty years away, had not the wartime quest for an atomic bomb meant a single-minded and almost desperate concentration of research and industrial effort which by normal standards would have been quite uneconomic. And so, oddly enough, the prosperity which the world is about to enjoy is to some extent due to technological progress directly attributable to the holocaust of the last war. With this long jump in scientific knowledge and technique, nothing stands in the way of a development which may result in energy being available at a price very much lower than it is now, and this may make oil less competitive than it is at present.

Some people have said that with the higher standard of living resulting from this cheap energy, oil consumption would rise even if it may then be once more used mainly for specialized purposes. For instance, if the standard of living rises as a result of 'more power to the elbow', more people will be able to run motor cars and use more petrol – this looks very likely, but are we so sure that, say, very cheap electricity might not call forth new methods of providing power to moving vehicles? Is it quite impossible that in such circumstances the art of storing electricity by some form of battery might be further developed? Is it ordained that the only satisfactory and economic method is to explode or burn the fuel right under your bonnet? I don't know, but 'there is no harm in asking'.

Be that as it may, I for one am not a defeatist as regards the future of oil. What I do say, however, is that the spell of oil as the providential and universal provider is now broken, and that a policy of conserving oil for future generations is no longer a public need; indeed it is probably no longer a sound business proposition. I am almost certain that the oil companies have realized that oil should now be sold in rapidly increasing quantities whilst the selling is good.

Such increase solves some problems and, as you would expect, creates others. The other day a civil servant asked me what my choice would be if I had to express the main problem of the oil industry in a single word. I had no hesitation in replying: INVESTMENT. That is to say, the growth in turnover, and the shift of accent towards fuel oil, requires the industry not only to multiply its oil wells, tankers and storage installations but to replace some of the refineries by others of a new type of design. Where can or should the money come from? The industry maintains that most of the new capital can only come from current earnings. That means in cold fact that the difference between costs and the sum total of returns from sales, that is to say the gross profits, must remain as high as it has been of late, or should become higher still.

Experts in America have shown that the oil industry has been in the habit of finding 85% of the money required for new plant from current profits. They therefore assumed that this would continue to be so. But what is happening now in Europe and the East is: the requirements of capital are becoming suddenly much bigger because the industry starts to grow faster than it used to do, and consequently profits on *today's* turnover would have to increase a great deal to finance *tomorrow's* business. We are told that it is a good thing that the greater part of the profits is being 'ploughed back'. This term, with its agricultural flavour of fresh air and virtuous thrift, does more to confuse than to enlighten the public. These profits, before they can be ploughed back, have first to be made, and the industry which makes them subjects all others to a kind of compulsory saving. If there was not so much of the profit reinvested, profits of this magnitude would never be made since they could not be distributed without attracting unfavourable attention. Most likely prices would be lower, that is to say other industries would make bigger profits and individuals might have more money to put where they liked. Admittedly the oil industry as a whole needs more capital now, and we are told that the so-called 'capital market' is not sufficiently wide today to provide this new capital; that means, there are not enough other savings available to satisfy the needs of the oil industry. This is probably true now, because of high taxation and no one can blame an industry for getting the money where it can, especially if it is put to good use. Yet this practice has its dangers. Some industries, of which oil is only one, are so organised as to be able to be judges in their own case. If they take to self-financing, by setting their profit target sufficiently high, the very principle of a free-flowing capital market will be affected. This capital market is normally fed by savings of many individuals and by profits of *all* enterprises. If it becomes a general habit of 'strong' industries to pre-empt a large slice of what is available, we are getting into a vicious circle: because the capital market is insufficient, some industries intercept the money before it gets there; thus the capital market gets weaker still and so ad infinitum.

Anyhow, the picture of the oil industry as I have drawn it here tonight is one of great promise: we have seen continuous increase in demand, which lately has gathered a still greater momentum. The oil industry – on a world-wide basis – has got what it takes to meet this demand; indeed, almost one hundred years after the sinking of the first crude oil wells, the industry is as virile, as resourceful and as elastic as ever. I have mentioned the deferred threat of new sources of energy. It may be more than a threat in due course; today it is, I believe, but a spur to rapid advance by more generous use of our oil reserves. The difficulties, such as they are, seem to result from oil's prominence in international and national affairs. As we have seen, much of the world is split into producer countries and consumer countries; also, just because oil has become so important, public opinion cannot be indifferent to the question of oil company policies.

The oil companies have become so big, and there are so few of them, that the public sometimes feels uneasy about almost anything including prices and profits as they result from the type of competition that exists between the oil companies.

The only remedy for such distrust is to take the public into one's confidence. The basic position must be realized: that the oil companies are today – and have to be – part of a private enterprise system and at the same time trustees of public interest.

4   A Turning Point

This piece, written in October 1957, contains another clear and sane statement on the nature of oil industry competition and the reality of the 'cartel'. His description of the producers finding themselves in 'hypothetical agreement' to limit exports in exchange for income stability and the comment on the 'enlightened paternalism' engineered by the companies is, in retrospect, a felicitous commentary on what has always been a love-hate relationship. The significance of the 'independents' entering the production scene is, in this post-Suez article, given early notice and there is, in the penultimate paragraph, a timely warning that an OPEC (as it later turned out) might be formed.

The Era of Equilibrium

Any analysis of the present state of the international oil industry has to take as its starting point the developments of 1944–46. When the people directly concerned in industry and within the governments looked ahead beyond the period of wartime controls they appear to have been unanimous in their intentions to avoid the troubles which gave the years after the First World War their characteristics of turbulence and allround cussedness. Although in 1944 it was too early to say exactly what the future relative positions of the American oil companies and their British (and British–Dutch) competitors and colleagues were to be, those in control on both sides of the Atlantic were determined to learn from history, and the Anglo-American oil treaties of 1944 and 1945 were in fact merely an *agreement not to disagree*. That is to say the signatories (even of the emasculated 1945 version which at the time was likened to a Japanese kimono covering everything but touching nothing) provided in the International Petroleum Commission a machinery of consultation which was designed to keep things on the rails. In the end the U.S. Senate failed to ratify the Treaty, the Senate being then more than now averse to foreign entanglements of an unspecified nature. True enough in its tendency to be 'agin it' the Senate was appropriately spurred by Texan and other interests who might even have accepted an international oil peace, but thought that anything of this nature would give Washington in the end control of the domestic oil industry as well. Some people are more afraid of their own government than of the foreigners.

Failing a machinery on the governmental level, and nature abhorring a vacuum, the oil companies had to establish a modus vivendi as best they could. The problem on hand was then the disequilibrium of the various operators: there were those who had been farsighted (or lucky) enough to have gone into the Middle East in a big way, into an area which at that time was beginning to show its stupendous possibilities: Anglo-Iranian, Standard of California, and the Texas Co. None of these three had sufficient facilities, i.e., refineries and markets, to take care of the potential quantities to be derived from their concessions; such facilities were in the hands of the traditional international majors: Jersey-Standard, Socony-Vacuum and Royal Dutch/Shell, who had, compared with their own size, only minor production in the Middle East but who, put together, controlled more than half of the oil sales outside the U.S.A.

The real oil settlement of the late forties consisted of re-establishing an equilibrium between the oil-haves/market-have-nots, and their opposite numbers, whose deficiencies and whose strength were altogether complementary (and incidentally there were British and American companies on either side of the fence). This was the period when the die of the postwar oil set-up was cast; almost simultaneously the several plus and minus positions were settled. The then owners of Aramco (California-Standard and Texas) sold 30% of its shares to Jersey and 10% to Socony; Anglo-Iranian concluded a substantial long-term sale of crude oil to Jersey and Socony, and Gulf did the same (though on somewhat different terms) to Shell, who also undertook to offtake from Anglo-Iranian. It is essential to realise that all this did not involve a rigid agreement on markets and prices but it reduced the likelihood of, indeed the need for, violent changes in the markets and by doing so obviated inter alia a price-war which might otherwise have developed. If the original owners of Aramco and of A.I.O.C. or Gulf had had to squeeze into the markets the quantities they had to sell they would have had to change the distribution positions in *each* country which would have required a colossal effort for which they would have had to find the money at a time when they would have been on their own in providing finance for the development of their resource, of pipelines, tankers and all.

To call the resulting set-up a Cartel is disingenuous. The simple fact is that such reasonably stable equilibrium made a cartel unnecessary. To say that there were some Secret Restrictive Covenants without which the situation would not have been as orderly as it was is on a par with a statement that there is no visible revolutionary movement in England or Switzerland because of its Repression By The Ruling Classes – whereas the truth is that there just is no revolutionary *situation*.

Marginal Competition

All this does not mean that there was no competition outside the U.S.A. – indeed there was a lot of it, but in the nature of things it remained on the fringes. Although the main positions were undisputed there was lively, and sometimes heavy, skirmishing between the trenches, guerilla warfare of sorts. There was competition for the custom of the few non-integrated refiners and under-the-counter benefits became customary. Sale of crude being rather remunerative benefits tended to be capitalised; independent refiners and marketers soon realised that it was easier for the supplier to buy them out altogether at an inflated price than to go *too* far in underquoting his own posted prices, a practice which might have brought down the whole edifice; these refiners and marketers being thus worth more dead than alive sold themselves at a price. Another characteristic of these competitive contests was the octane inflation in Europe which gave the motorists a quality of fuel far above the capacity of their cars to benefit therefrom; and on the lunatic fringe of the competitive field we saw a craze for additives designed to improve motor fuels which, however sound their conception might have been on the testbed, fell inside the margin of error of the average car, that most patient but therefore not very discriminating contraption.

But the biggest item in the field of competitive selling was the development of retail outlets especially for motor fuels. Since service stations have become the principal, almost the exclusive, means of selling gasoline, and since you must sell *some* gasoline to make the other products in the refinery, the value of a secure outlet for gasoline tends to be regarded as the key to everything else and it remains a commercial proposition even if it does not provide a genuine margin per se. Thus some of the companies have gone through what one would call 'Ordeal by Investment' even though this drive has not yet led to a major dislocation anywhere, except perhaps in Italy. A rapid increase in consumption may eventually justify even seemingly extravagant investments – in the oil industry today's madness is sometimes tomorrow's wisdom.

Sources of Supply

Whereas there was an overall equilibrium of productive crude oil capacity and of marketing facilities this, in itself, did not solve the problem of allocating the existing outlets to the several alternative sources of supply, i.e., the various Middle East countries and Venezuela. True enough, some of the decisions were taken for, not by, the operating companies – in the case of Venezuela it was price structure and dollar problem, in the case of Iraq the need for costly pipelines and port facilities which took long to build, and in the case of Iran the folly of the Moussadek experiment. Yet there remained to the oil companies a certain amount of discretion and its use was partly determined by the fact that most of the companies – and since the Iranian Consortium was set up all of them – had more than one source of supply in the Middle East.

Thus, not only was there the overall equilibrium mentioned earlier on but there was a degree of flexibility which made for a smooth development. Consequently the underlying competition of the several producing *countries* never came to the fore and, incidentally, never had an opportunity of expressing itself in the price of crude. The set-up as it emerged was in its results tantamount to a hypothetical agreement of the producer countries to limit reasonably 'their' exports in exchange for there being no pressure on prices as used for the purpose of the 50/50 profit-sharing agreements.

Although there is no doubt that this system of enlightened paternalism did work, it is perhaps a pity that the method of allocation among the producer countries of the quantities required has made it most unlikely that they could fully appreciate how difficult it is in normal times to find markets for their oil and to realise that they are in fact competing with a number of other sources of supply.

How can they understand the true relation of supply and demand in a world where there is much more readily available oil than there is current use for it, when they are confronted only with a number of oil companies anxious to get a concession (or, for that matter, with one or two who don't want to lose one)?

Further Outlook: Unsettled

All these conditions of reasonable saturation of the contestants are now likely to change – it is easy to perceive the beginnings of such changes but very difficult to see and say, where it will all end.

These are the portents:

(1) As an aftermath of the Suez crisis, and as its ironical postscript, the oil world is plagued by disproportionately high stocks of almost everything – the Horrors of Peace, as it were. If by a slight general recession, and even without it, the spectacular rate of increase of consumption of the post-war years, which our house mathematicians have been pleased to project into the future, should not be maintained, the fight for gallonage may become more serious than it has been of late.

(2) Some of the international companies who, ten years ago, were quite content to see the bulk of their crude oil refined and marketed in the Eastern Hemisphere by other groups, have lately given the impression of wanting to

quicken the pace of the development of their own outlets. A tendency which should become yet more pronounced should the restrictions imposed on imports into the U.S.A. become a permanent feature.

(3) These events are, however, dwarfed by another: for the first time since the war new participants in the field of large-scale crude oil production are appearing on the scene and they are likely to do so in a manner calculated thoroughly to upset a situation to which we have grown accustomed. There had been before, to be sure, some potential irritants, such as the Iraqi and Iranian royalty oils, some of the oil of the additional members of the Iranian Consortium and that of Neutral Zone, but by the logic of things the royalty oil never appeared on the market and all the others together did not amount to much. The position will be different though if there should be really large-scale production in Africa and if there should be a substantial production in Iran outside the Consortium area. Very little of that 'new' production would find a ready-made market, there are this time no organisations short of oil by which it could be absorbed 'without tears'. There is another new element: neither the joint Italo-Iranian company nor the Iranians alone, should they in fact develop the Qum field and manage to have a pipeline built to the Mediterranean, will be able to sit back and wait until they can slowly infiltrate into the markets; and – looking at the way this development is financed – will not, once the initial effort has been made for drilling and transportation, the Sahara oil need to be sold then and there? And will the same not apply to those American minor majors who venture into Libya and elsewhere and possibly also to some of the Venezuelan crude produced from concessions acquired by the payment of inordinately high premiums?

We must get used to the fact that new oil – thinking back twenty-five years one might almost call it 'hot oil' – presses on the market in a manner quite different from developed reserves in the hands of 'balanced' companies.

Various things may happen: there is the possibility that

some of the 'new' oil will preempt certain markets in which it may acquire some preference, thus shutting out (and back) other supplying areas. Some governmental influence may be wielded to facilitate the entry of contestants who may otherwise not be able to make the grade.

Alternatively – or simultaneously – *the pressure of some sellers who have nothing to lose elsewhere* may in the end express itself in the price of oil – one could envisage a situation in which even some of the established producer countries, far from trying to get a higher price, would start calling to be allowed to be competitive or even to be 'more equal than others'. This different type of competition may result in prices which could no longer be rationalized by world-market-price theories.

Obviously this change of the set-up in the sphere of production might have its repercussions further down the line – there would be a tendency towards integration with new partners and additional distribution networks may be developed from scratch. This may be made easier by a still more marked pre-eminence of fuel oils whose sale needs less elaborate facilities than does that of gasoline.

Finally the wheel may turn full circle and the producer countries might one day call for an international order for oil as a means of securing for themselves an 'equitable' share in the markets. The history of the oil industry in America has taught us that when in the early twenties there was a scare of possible oil shortages nobody in the industry was very interested in oil conservation; but when the wind started to blow the other way and the over-production of 1927 began to concentrate the operators' minds, there was no delay in calling in the authorities to 'regulate' the situation.

In spite of the difficulties ahead there is no reason to be despondent; the ability of the oil industry to take even rapid changes in its stride is one of the sources of its strength. Our industry got where it is today, not by just making the stuff available, but by aggressive selling. A shake-up now may put it into a better position to get its share in meeting the energy demand to come. Oil should, even on a new basis, still attract the capital it needs.

5 Topical Problems

Frankel formed his company, Petroleum Economics Limited, in 1955. From the outset his clients were supplied with a two-monthly commentary on relevant market developments in the industry and from May 1957 this was introduced by a section entitled 'Topical Problems'. These 'topicals', as Paul Frankel refers to them, are a continuous commentary (from 1984 to 1988 they were written quarterly instead of every two months) on the oil industry and its problems over a period of over thirty years, given a unique consistency in that Frankel himself wrote every one of them. Several are reproduced in this volume. This one, written at the end of 1959, looks back at the 1950s and forward to the 1960s. His suggestion that we might have seen the 'highwater mark' of the tide of producer country income demands, a reasonable enough proposition in terms of supply/demand at the time, was not, of course, to be borne out in practice. OPEC was formed less than a year later.

The end of the '50s and the beginning of the '60s is an appropriate moment for a brief review of the prospects of the oil industry in the light of past experience: there can be no problem more topical on 31st December 1959.

Looking back at the last decade, we see a tremendous progress in the demand for oil and natural gas, the increase being more marked in the Eastern Hemisphere and in Latin America than in North America. This is due to the first-mentioned areas now entering a stage of industrial civilisation which the U.S.A. had reached between the first and second World Wars. Consequently it is perhaps a justified assumption that in the next decade the highest rate of increase will no longer be in the Western World but further east, including the Soviet orbit.

Another element in the supply picture is the advent of

November/December 1959.

natural gas as a major factor. In the U.S.A., the construction of long-distance pipelines has extended the use of natural gas from the South-West where it mostly originates to almost all the centres of population, and Canadian gas found in the West (Alberta) is now available not only in Eastern Canada but also in California. In Italy, where in the Po Valley for once natural gas was found near centres of consumption, it has made a great impact during the last decade. Finally the development of the prolific gasfield of Lacq in the south-west of France has shown how great the influence of natural gas can be if it is found in areas within reach of economic transport to consuming areas, even if the location is not as favourable as it is in Italy. On the other hand entirely new developments, such as the technique of tanker transportation of methane, has widened the area in which natural gas can make itself felt.

Another spectacular event of the '50s was the extent to which fuel oil could make inroads in the supply of energy even in countries such as the U.K. and Germany whose energy supply is traditionally based on coal. This is due mainly to the increase in the price of coal during the last decade at a time when the pressure of oil supplies kept fuel oil prices down, whereas the incentive for the oil companies to increase their sales was kept up for the same reasons and also to the greater ease of handling oil or natural gas as compared with coal.

To mention this 'pressure of supplies' is to deal with the main characteristic of the last few years and possibly with the most critical problem of the next decade: the discovery of oilfields of unprecedented magnitude and comparative ease of exploitation in the Middle East, the great event of the '40s, became fully effective only in the '50s. The discovery of entirely new and partly most promising areas in Africa was one more step in the same direction. All this has, at least in time of peace, done away with the recurrent fears of a worldwide oil shortage. It has furthermore made it possible for the oil companies involved to meet rising demand with low-cost oil, and the substantial margins of profit resulting therefrom provided the financial basis for investments on a very large scale. It was on the other hand inevitable that the governments of the producing countries strove to increase their participation in the margins. The conception that the difference between the cost of producing oil and the Posted Price should be shared fifty-fifty between operating company and producer government is only just over ten years old, and although at the existing Cost/Posted Price ratio it provided a stupendous increase in the producer government income per barrel of oil, it looked at the end of the decade as if this was a minimum upon which producer country governments could hope to improve considerably. But by now we might have seen the highwater mark of that tide: the fact that so much new oil has been found and is being developed has reduced the bargaining powers of each producing country. These countries are increasingly faced with the facts of life and will no longer be in a position to make higher and higher claims to be satisfied out of a seemingly bottomless purse of the big oil companies.

Will the increasing competition for a share in markets which has grown much more slowly than has supply lead to a period of extreme and ruthless competition – fought by means of low prices and government-imposed preferences? Or will there be a more comprehensive settlement for the various conflicting interests on a rational basis?

6 Topical Problems

This 'topical' records the creation of OPEC in September 1960. The background of over-supply and weak markets was referred to in the previous piece. Pressure on the market intensified through 1960 until in August Esso (it was transformed into Exxon in 1972) reduced the posted price of Arabian Light at Ras Tanura by 14 cents per barrel. The Esso announcement was the trigger that enabled Perez Alfonso of Venezuela and Abdulla Tariki of Saudi Arabia to carry out their plan – conceived in the previous year at the Arab Petroleum Congress – to set up the meeting in Baghdad that led to the creation of the Organization of the Petroleum Exporting Countries on 14 September 1960. Frankel's final observation that 'one can probably expect that there will be a period of reflection and immobility' was borne out by events.

In our last issue we dealt with the problem of Posted Prices and the significance of their reduction which had then taken place. Since then a number of important developments have begun to show what may be the trend in the next few years.

It can now be assumed that the endeavour of some international oil companies, in the first instance Standard of New Jersey, to bring Posted Prices nearer to the real level of the current market value has actually failed. Some oil companies had reduced their prices by only about half the reduction carried out by Esso – and the actual Esso reduction was already much less than they had intended to effect originally. All companies who post prices including Esso ended up in September with comparatively small reductions. It should be added that there has been no change in Venezuelan Posted Prices at all, and that the countries, especially in the Middle East, in which Posted Prices have a significance within the system of profit-sharing have so far stated that they will not

September/October 1960.

recognise even the small reductions for the calculation of the 50/50 profit-sharing system. The real significance, however, of the endeavour to reduce Posted Prices lies in the fact that the unilateral action of the oil companies in reducing Posted Prices which lead to producer country revenues being reduced, has resulted in a closing of the ranks of most of the relevant exporting countries in both Hemispheres into an organisation called the Organisation of Petroleum Exporting Countries (O.P.E.C.).

The very conception of O.P.E.C. is the result of a number of changes which have taken place during the last few years. These are:

(1) The loss of control over the international markets by the big international oil companies.
(2) The repercussions of the U.S. governmental import control.
(3) The fact that more people now understand the direct relationship of the volume of supply to current demand and the price of oil.

The thinking which has resulted in the O.P.E.C. conception is based on an acknowledgment of the facts just listed under (1) and (2), and all this may in certain circumstances represent a genuine progress towards a realistic appraisal of the situation. It remains to be seen whether the united front presented at the moment can be maintained when the countries come face to face with the practical problems and realize the diversity of the position of Venezuela with less than 18 years reserves and the Middle Eastern states with an average of 108 years of reserves at the 1959 level of production. On the other hand, if the producer countries should succeed in finding a common denominator and become so impressed by the apparent power which they could have in forcing high prices on the world by restricting supplies that they should try to misuse this power, the repercussions could be serious on all sides.

In that case they would strengthen the already existing tendencies in consumer countries to protect other and indigenous sources of supply. Also consumer interests may try to break up a structure such as O.P.E.C. by promising to purchase from suppliers which remain outside O.P.E.C. or would be prepared to leave it. Finally, any threat of oil shortage or high prices for oil would encourage long-range investments in alternative sources of energy such as tar sands, atomic fission, etc.

In the circumstances one can probably expect that there will be a period of reflection and immobility and much depends on the studies which will be made from now on by producer countries, oil companies and consumer interests and on the endeavours to coordinate these interests.

7   Topical Problems

This 'topical' is an early discussion of one of the core objectives of the OPEC producers, but whose negotiation was in practice postponed until the early 1970s. OPEC chose to fight on the price front first and the whole debate on participation versus nationalization and the nature of participation did not take place until after OPEC had reformulated its objectives in Resolution XVI-90 in 1968. Frankel was, in the context of OPEC in 1961, well ahead of the game in discussing the concepts of participation at this early stage.

The most significant of the current developments which affect the international petroleum industry is undoubtedly the changing pattern of oil concessions in producing countries such as the Middle East, Venezuela, Libya etc. In the history of oil concessions in these countries one can recognise a definite trend towards greater participation and influence of producer countries. Originally most concessions – apart from some flat fees payable for the area of the concession, sometimes called 'dead rent', or similar initial payments – consisted in a royalty in the strict sense of the word, i.e. payment (in kind or in money) per barrel of crude oil produced. In this way the government granting the concession was interested in the volume of production and in some cases in the price of the crude, but not directly in the profit made, i.e. in the difference between cost and price. It must not be forgotten that in the period before the Second World War crude oil prices were comparatively low and producer countries did not want to be too dependent on the actual profits made (if any). This changed after the war, when, in the Middle East and to some extent in Venezuela, prices were and remained for some time very high in comparison with the physical cost of crude oil

September/October 1961.

production. It was at that stage that the conception of profit-sharing between oil company and producer country was developed. From then onwards it became inevitable that the producer country governments, who had begun to understand more clearly the problems involved, were of the opinion that being 'partners' they should be consulted by the concessionary company about matters affecting their share in the profit, especially on price policy.

At about the same time some producer country governments began also to ask questions about the profits made by oil companies at the subsequent stages of the industry which take place beyond the borders of the producing countries, i.e. transport, refining and marketing. The 'partnership' at one stage of their operations of an integrated oil company with third parties does create a genuine problem, in fact the system of 'posted prices' in Venezuela and in the Middle East as we now know it was developed after the war in order to give the producer country a certain assurance that the oil companies would not supply their own affiliates abroad at prices which were *lower* than 'world market prices'. Greater competition in consuming countries and the resistance of producer country governments to seeing their share in the 'profits' diminish have resulted in the opposite phenomenon, i.e. posted prices are today *higher* than those at which crude oil can be sold to non-integrated buyers and higher than what some of the affiliated companies can afford to pay for the crude oil. In these circumstances the suggestion of some producer countries that they would like to participate in the subsequent stages of the industry does not make much sense because money is lost on some of these activities when crude oil is put in at the 'posted price'. Anyhow, except in the case of newly formed companies, it would be very difficult (and for the producer country government extremely costly) to acquire a share in the existing integrated operations of the big international oil companies.

It must be assumed, however, that for economic and political prestige reasons the demand for actual participation by shareholding of these governments in the producing *and* in the subsequent stages of the industry will remain to be made. One way of identifying the producer country with the subsequent operations in other countries was devised in the concession of the Japanese-owned Arabian Oil Company, but as the Japanese company was not integrated, i.e. had no refineries and marketing affiliates, it is not clear yet in what way the terms of that concession will become operative.

Some oil companies, on the other hand, have recently studied the idea of paying taxes in the producer country not on a posted price, which may be quite unrealistic, but on what is called 'actual realisations'. This tendency has obviously something in common with that of the producer country governments which we have just mentioned, and it is quite possible that out of all these ideas a new form of cooperation will emerge. It is no longer possible for the oil companies just to maintain that the '50/50' profit-sharing is a complete, and the only, solution, because all depends on the methods by which the profits to be shared are being assessed.

Although in theory the system of profit-sharing by realisation is logical, it is extremely difficult to agree on an equitable way of calculation. That it is not altogether impossible has been proved by the comparatively successful execution of the supply contract between Gulf Oil and Shell for Kuwait crude. Whether one can find a method which would be understood and be acceptable to a number of governments remains to be seen. It is clear that if the value of the crude at the border of the exporting country can be assessed in a fair and workmanlike manner, the problem of large-scale participation of producer countries in the subsequent stages of the industry does not arise, because the price of the crude would then express its real value. This, however, would not exclude that producer countries could, if they so wish, acquire positions in transport, refining or marketing abroad, i.e. to become, on a smaller scale at first, integrated oil companies, either alone or in cooperation and partnership with existing oil companies. In this respect the most recent formula evolved in the Kuwait offshore concession obtained by Shell is perhaps an indication of things to come; it does open the door to a gradual infiltration in the sphere of actual business by the concessionary government and in this respect it has some elements in common with the earlier concessions obtained from Iran by E.N.I. (SIRIP) and by PanAmerican (IPAC).

These developments, although they are less spectacular than an open conflict such as has arisen between the Iraq government and the Iraq Petroleum Company, are yet the more interesting and in the long run more relevant ones. It can perhaps be expected that a new conception will develop by mutual consent.

8 Oil: The Facts of Life

In September 1962 Paul Frankel published Oil: The Facts of Life. *This was a short essay in paperback, a total of thirty-eight pages, in seven sections. The last three sections – 'Oil Policies', 'Producers' Lobby' and 'Conclusions' – are reprinted in an edited version here. They are still worth reading since the fundamental principles and perceptions on which they are based are as relevant today as they were in 1962. It should give European politicians and administrators (if any today can remember what they were discussing in the 1960s) pause for thought to be reminded that their concern* should *have been with having to pay 'too high a price for crude'; few of them were worrying about that for another ten years. It might give today's planners in OPEC countries pause for thought when they read what Frankel was saying twenty-five years ago about prorationing (the word used then for running a cartel) or about the implications of downstream investment and integration. Some of us perhaps may be less impressed by Frankel's tentative sketches, in the 'Conclusions', for solving the ever-present differences of objective between oil consumers and oil producers but grappling with this problem has been one of his continuing preoccupations. Already in* Essentials of Petroleum *(as he repeats here) he saw the need for strategic guidance and supervision over oil policy by consumer governments; why not international strategic guidance as well? This is a theme to which Frankel will return. And in the last paragraph of the essay he invents the phrase which by implication he applied to himself – and would like others to apply to him – the Common Carrier of Common Sense.*

Oil Policies

The concern of national and international authorities with oil is only natural: it is the main source of income for some and now an indispensable supplier of energy for all, also it is still

Weidenfeld & Nicolson, London, 1962.

the biggest item in international trade both in terms of volume
and value.

There are clearly three types of countries whose focus on oil
depends on their respective positions:

(1) Those who are to a great extent self-sufficient, i.e. can
    provide adequately, though not always most econo-
    mically, for their own needs and whose export interests (if
    any) are not vital – in this category fall the USA and the
    USSR and in some respects Canada, Mexico and now the
    Argentine.

(2) Those who have a production much larger than their own
    (quite often insignificant) domestic demand and depend
    on large-scale and regular exports: there we find Vene-
    zuela and all the producing countries of the Middle East
    and now of North Africa.

(3) Those areas – which include all of Western Europe and
    Japan – with a considerable and still rapidly growing
    demand and virtually no indigenous oil resources. There
    are many other countries in the same category, but
    amongst them only Australia and New Zealand have a
    high per capita oil consumption.

The *oil policies of the USA* are well-known: pressing demand
for oil – the US was the first country to be 'americanised' –
and lively competition among hosts of inventive *entrepreneurs*
made for a rapid opening up of resources which, as we now
know, are, as compared with other areas, not particularly
ample. The somewhat wasteful methods of American oilfield
development, the high cost of their particular brand of proration, and the elevated level of profitability to which oil
producers have become accustomed and which they manage
to maintain by way of a singularly effective political lobby,
have made it necessary for US oil to become detached from
the rest of the oil world. One can have different views as to the
most adequate protective methods to be applied and the present system of mandatory import controls may become subject
to alterations, but it is unlikely that there will be a change of
heart as long as the claims of domestic politics are being
underpinned by serious concern about national security or
even about their status as a first-rate power.

Anyhow, the Americans have solved after a fashion the problem of the different levels of delivered cost of indigenous and foreign oil – indeed one can gauge the incidence of this difference by adding up the import tax of 10.5 cents per barrel and the amounts allowed in some form or other for import quotas, which inland refineries who can't use them pass on to coastal refineries who can, amounts which now run to more than $1.00 per barrel. The USA still consumes more than two fifths of the world total and the fact that the biggest market for oil in the world has cut itself off by strictly governmental measures from free interchange between areas, must have serious repercussions elsewhere, to put it mildly.

However much the *position of the USSR* is being taken for granted, it is worth analyzing. We don't know much about crude oil production cost in the Soviet orbit, but it is very likely that they too have a very wide range of costs and seem to be content to average them out in some way or other. Looking at some of the returns they are getting from export sales one wonders whether they relate them by any chance to the lowest-cost oil they have, ignoring the fact that they replace it at home with high-cost oil, and whether it would not be more rational to abandon such high-cost production and to stop exporting. Such considerations would be outweighed, however, by some overriding desire to acquire foreign exchange by exporting a readily marketable commodity.

It is worth remembering that international oil trade is for the USSR a one-way operation since they are, by definition, impervious to inroads other people's (cheap) oil might make in their markets. In the circumstances the Moscow spokesmen who complain about restrictions which are or might be imposed on their supplies to some countries could be usefully reminded of the classical retort to the propaganda against capital punishment: *que messieurs les assassins commencent.*

Canadian oil policy is peculiar inasmuch as she is in a position of basic self-sufficiency – with most of her oil in the cost range of USA production. Her oilfields are, however, eccentrically located to her own consuming areas and this makes Canada an exporter *and* an importer. Canada is beginning to solve her problems by what is called a National Oil Policy which, if the operators stay within walking distance of its general precepts, could make it unnecessary to apply hard and fast rules; we shall hear more about this presently.

The idea that there can be such a thing as a rational *policy for oil producing and exporting countries* is a comparatively recent one. Perversely, low-cost oil in quantities has in this century been found only in countries which, for climatical, historical and demographic reasons were not previously industrialized and thus – past glories notwithstanding – did not have the business and governmental structure which could carry as heavy a load as was the emerging oil industry. Thus the stage was set by the oil companies only, and it was they who established originally a pattern of relationship which, however iniquitous it may look in retrospect, provided hitherto un-dreamed of benefits to the host countries.

It is normal that the loudest complaints about one's stan-dard of living do not come from people (or peoples) in an acute state of poverty, but are voiced only when things have improved and one has the leisure to contemplate, and the educational standard to envisage, how much there is still to be desired. This is the stage at which most of the oil countries are now, and the oil companies are justified in feeling that the wealth their very activities have unlocked for the inhabitants and their very educational effort have created some of the difficulties with which they are now faced.

The so-called 'host governments' are confronted with a limited number of powerful and sophisticated companies who, though competing with one another, will not easily act in a way which would prejudice the basic tenets which they inevit-ably have in common – dog bites dog, but dog doesn't eat dog; consequently it is obvious that oil producer country govern-ments should consider co-ordinating their strategies. The following chapter will cover the significance of such a move, its prospects and limitations.

Before reaching that stage it is however necessary to look at *consumer country oil policies*, the very idea of which is of recent vintage. For some time there seemed to be no need for such a policy, since in a state of ample supply – sufficient to meet demand within the foreseeable future, allowing for even the more ambitious forecasts – the pressure of competition among the potential suppliers would assure the availability of all that is required and on the right terms too. Taking this line, European concern was during the last few years directed mainly towards harmonization of their own indigenous energy supplies with imported energy whose price was in some instances lower than could be matched by most of European coal. In this coal/oil relationship we recognize some of the elements of the very problem of high-cost and low-cost oils and their co-existence. In line with that pattern, European coal producers and the governments which are responsible for their wellbeing were disturbed by the 'competitive excesses' leading sometimes to 'unduly low' prices for fuel oil, rather than concerned with the fundamentals of the situation.

To come to grips with the real problems of European energy policies progress had to be made in two directions:

(1) It had to be realized that – contrary to its deeply ingrained historical experience – Europe was no longer an exporter of energy, nor was it any more reasonably self-sufficient, but that it had come to depend, heavily and on an increasing scale, on foreign supplies. To appreciate this fact and to recast one's thinking altogether was as hard as must have been the acceptance of Copernicus' newfangled idea that the earth was not after all the centre of the universe.

(2) The other illusion to be shed was that the competitive system alone could be relied upon as the arbiter of the terms of trade in energy. This conception has been the mainstay of oil company propaganda for some considerable time and the market trends seemed to confirm the tale. That things are not as simple, however, is borne out by the price alignment of low-cost oil to the higher-cost variety which obtained until only a few years ago and, yet more poignantly, by the current tendencies of producer country governments to band together to keep oil prices high. The conception that oil companies could be relied upon to fight the consumer countries' battles could turn out to be an illusion: if it comes to it oil companies, understandably, nay inevitably, will lean towards the side which can bring to bear upon them the more concentrated and immediate pressure; it is not difficult to see where such power lies in the first instance.

Taking both arguments together, one can sum up that the concern of Europe in matters energy in the years to come will not be so much the bother caused by low-priced fuel oil, which competes with coal, but the danger of having to pay too high a price for the crude oil it has to buy. It is only fairly recently that the European problem – which is by and large that of other areas bereft of indigenous energy – has been formulated in a workmanlike manner. This is to a great extent due to the perceptiveness (and, incidentally, to the courage) of some of the European civil servants in Brussels and Luxembourg. It is now well established that in the energy field consumer countries can no longer sit back and rely on things coming out all right in the end. All the other participants in the game whose positions and postures I have just listed have had their claims armour-plated by their governments and some have also had some weapons with an aggressive potential built in. It is obvious that Europe could not remain indefinitely a *locus minoris resistentiae*.

Producers' Lobby

OPEC was formed as a direct result of the endeavour of some oil companies to cut their Posted Prices for Middle East crude oils substantially and of the decision of all of them, reached after some prolonged skirmishing in August 1960, to reduce them moderately. Whereas this shock released even in the more cautious governments of producer countries the spirit of self-preservation, the idea of a close alignment of the several oil producing countries' governments had been developed earlier on and had in 1959 led to a plan proposed by Dr Perez Alfonso of Venezuela and Sheik Abdullah Tariki of Saudi Arabia. The initiative seems to have come from the Venezuelans, who, understandably enough, were concerned about their competitive position: they had not succeeded in getting a cast-iron privileged position by way of 'country quota' which they had hoped the US mandatory import system would provide for them and, with tanker freights at what looked like a permanently low level, they had lost part of the location advantages they have over the Middle East, which is more distant from most of the centres of oil consumption.

In respect of oil and gas reserves and of cost, Venezuela is in between the USA and the Middle East levels, mostly rather nearer to the former than to the latter, thus they will tend to take the same route as do US producers: if called upon to decide whether they want to produce as much as they can do within the limits of sound production techniques or would rather aim at the highest price per barrel they can get, they will first try to have both, but, if that is not possible, will sacrifice quantity to safeguard price.

This must have been the spirit of Dr Perez Alfonso's approach to Sheik Tariki and in all honesty he would have had to say: we here exploit our proved reserves at the rate of sixteen or seventeen years and our costs are quite high; your oil is being lifted at a rate which would exhaust reserves in more than a hundred years and you have much lower costs; let us both be reasonable and establish a uniform pattern for limiting our production to get the best prices.

As we have seen earlier on, a low-cost producer with large reserves, ie reserves he cannot expect to see the end of at any reasonable reckoning, would in the absence of overriding (or ulterior) correctives be perfectly justified in opting within reason for maximum output, never mind the price. If he is the only one to be in this position his price will still not fall below a certain floor, likely to be well above his own cost, and he might be better off all round. If therefore the prospects for a straight Perez–Tariki axis were none too good, the very emergence of such plan was symptomatic: it proved that the facts of life had come to be understood, ie that in the circumstances *the real problem of international oil was one of quantity*. Prices, and the margin between them and the physical cost, planning for investment and the use of facilities once they exist, are determined by *the fundamental decision whose oil reserves are going to be developed to what extent and in which order*.

If the preceding analysis is not altogether wrong, the order of priorities has not been established and is not likely to be established by the criteria of cost alone. I have also shown that the pressure of low-cost oil *has* made itself felt, but it has generated countervailing tendencies liable to check it before it went too far. All these tendencies fall into the same category: the mandatory import regulations in the USA, the revulsion of producer country governments against a reduction of their participation in the profits and the opposition of competing purveyors of energy, such as the coalmining industry.

What are the means now at hand to establish deliberately an equilibrium which is apparently what is desired by all concerned? A straightforward agreement among the oil operators, or at least among some of them, a cartel in short – which, without supervision, would be party *and* arbiter – is not only, as we have seen, impracticable on legal grounds, it would now be unacceptable morally and politically; it is a long way to Achnacarry.

Another approach would be an understanding among producer country governments, and endeavours to that effect have led to the formation of OPEC, an event of great significance. It was conceived in the spirit which had led to the Perez–Tariki dialogue, but it was actually brought into the world as a defensive weapon to protect the margin, the 'take' of the producer governments as it happened to be prior to August 1960.

When OPEC was formed its promoters behaved as if the millennium was in the offing: they had found the sovereign remedy of all ills, the philosopher's stone: if only all the major producing countries stuck together they could supply oil to the world at their terms and who was to deny them the lion's share? All that, however, far from being new, was in fact old hat: it has been known since time immemorial that an absolute monopolist had great power but that the problem was how to establish such watertight monopoly, and if one managed to establish it, it was equally hard to keep it in being.

The world was rightly startled when at the Second Arab Petroleum Congress of October 1960 Sheik Tariki read a paper on 'The Pricing of Crude Oil and Refined Products' in which he calmly stated that the oil companies had deprived the producer country governments over a period of seven years of $2,737,145,066 by not selling Middle East oil at a price in line with that of Venezuelan crude delivered to north-western Europe.

Sheik Tariki must have taken at its face value the *post festum* rationalization of the historical development of Middle East crude oil prices and accused the oil companies of not having adhered all along to a theoretical price schedule whose continued enforcement could have been based only on a rigid and all-embracing cartel.

In fact some people around OPEC realized that it would not be possible to establish and maintain the prices which they thought oil should fetch without controlling the level of output and there was much talk of international proration. To work out formulae to allocate quotas or to establish 'production allowables' was to bring to the fore the enormous difficulties which stand in the way of a deliberate and watertight solution of the problem, leaving aside the question that no formula could satisfy everybody. The USA example, so often quoted by advocates of international proration, shows how fallible and how primitive most of their regulations are even after thirty years of experience; also that, although some order has been established *within* the more prominent oil states, eg Texas, there is, the somewhat honorific Interstate Oil Compact notwithstanding, still not much of a co-ordination *among* the states. This alone shows that proration as a formalized system, however desirable it may be in the long run, is possible only if and as long as it is backed by an authority to which all involved are subject – to say that is to recognize that there is today no international authority in existence which could overcome the divergence of interests which stand in the way of a solution, an authority which could withstand the stress and strain which could easily develop within such a system if and when it was established.

It is not surprising that OPEC at an early stage dropped the hot subject of proration and seemed to turn its attention to an apparently more promising target: the return on capital invested in production of crude oil.

The margin between the Posted Prices and the physical cost plus the government's share according to the 50/50 pattern was still high enough to provide a massive return on capital invested locally in the task of producing oil and making it available at the point of export. This was the inescapable result of the setup described earlier on under which the price of some crude oils had become altogether detached from the actual cost involved in producing them. We have also seen how the pressures of competition made themselves felt in the first instance in the refining and marketing phases, with the result that practically all the money made by integrated companies appeared to be concentrated in the producing stage; this might in many cases have helped oil companies with their own tax problems – depletion allowance, tax credits and all – but it was certainly a great boon for the producer countries; it is interesting to see that the latter were slow in realizing that with the connivance of the companies they were really getting the best of the bargain and that the more these questions were made a subject of public discussion the more likely it became that their own outsize 'take' too would come under close scrutiny.

It becomes obvious that one could not altogether confine one's attention to the return on locally employed capital and had to take into account what happened in the industry as a whole, and seen from that angle the return on the total capital employed is not now exorbitant. That is to say the question is wrongly put in the context of the international oil industry as we now know it, the problem of 'capital locally employed' makes sense only to non-integrated newcomers. The magnitude of the return (to the successful operator) has in fact drawn the swarms of hopefuls into production all over the place. It would also make sense if production was to be entirely divorced from the subsequent phases, for instance by complete nationalization of the industry in the producing countries, their governments selling crude oil to foreign buyers by way of armslength transactions. It is unlikely that anyone would now be so bold as to advocate such a step, which would mean selling to the highest bidder without security of continuous outlet – supply and demand being what they are now, who could be sure that, far from the highest bid securing the purchase, it would not be the lowest offer which does it in the end?

Another version of the endeavour to improve on the traditional integrated setup run by oil companies would be the entry of producer country governments as *entrepreneurs* into foreign refining and marketing, either directly or through having a participation in the concession holder's integrated operations.

As to the former, there is no fundamental reason why

producer country governments should not put their money into oil refining and marketing rather than invest everything at home or buy other foreign shares, except that the latter operation means spreading the risk; also the very concentration of profits at the producing stage advocates against entry into refining and marketing which have become the poor relations, unless such step is inevitable in order to move one's own oil – *vide* the foregone conclusion to leave royalty oil with the concessionaires. Such forward integration could therefore be considered only if the terms of concessions (as they might develop) were one day to make producer countries the beneficial owners of oil for whose disposal they themselves and no partner were responsible; it may also make sense to establish foreign operations – pilot plants as it were – to test conditions so as to be able to verify the returns of one's major concessionaires.

An *ex gratia* participation by the host government in 'downstream' operations, ie transport, refining and marketing, of one's concession holder is an altogether different proposition. In its radical version it is based on the idea that the oil remains somehow the property of the government under whose jurisdiction it has been produced even after it has left that country. There is to be some sort of lien or mortgage to be attached to the oil involving a levy payable to the producing country apparently without the latter having participated in the effort required, including capital investment. Obviously if such terms could be agreed (and if agreed could be enforced, which is not the same thing) and if they applied only to some companies and not to others the margins earned by the consenting companies would be thinner than those of the others, the first-mentioned companies would not be competitive since they would in fact be paying more for their crude than others. To apply such terms to all companies at one and the same time would require a universal reshaping of concessions, which would hardly be practicable unless the whole basis of the government/company relationship is changed. I wonder incidentally whether the advocates of such system of 'integral profit sharing' have ever paused to think that on the strength of taxes now paid on illusory Posted Prices, with all subsequent margins pretty well eroded, they have got as near their grand target as they ever are likely to get.

Be that as it may, it is highly characteristic that in the first definite step OPEC has taken towards making its negotiating power felt – the resolutions resulting from its Fourth Conference – no fundamentally new trend of thought can be detected. In vain would one search the Resolutions for the brave-new-world look of earlier utterances, proration is not mentioned and there is no inkling of the erstwhile tendency towards extending the scope of producers into new spheres. Instead, in line with the lowest common denominator, principle has been sacrificed for the sake of expediency.

In fact the only remedy which the OPEC members seem to have been able to agree is the same old mixture, only the dosage is to be increased. All three demands – re-institution of Posted Prices in force prior to August 1960, consideration of (uniformly fixed, i.e. mostly increased) royalties as expenses not as tax payment, and abolition of the (nominal) discount hitherto allowed in some concessions for selling expenses – all three demands are in fact just gambits for a straight increase of the current government 'take'. The only new element in the resolutions proper is the tentative reference to the price of crude to be linked in future to the 'price of goods which the Member Countries need to import'. This is certainly an interesting aspect although it is not likely to be relevant on its own, but only as part of a more comprehensive framework of the means of balancing the interests of the several parties, a framework of which there is no trace in the Resolutions and only a faint vestige in the supporting Explanatory Notes.

At the back of the more farreaching as well as of the more pedestrian ideas of the future rôle of oil producing countries lies the conception of oil as being something intrinsically scarce and vitally necessary. Indeed if it were scarce there would be hardly any need to 'engineer' its price level; the fact that oil is needed, even that one cannot do without it, is not enough on its own – has anybody considered making bread dear just for that reason? Never mind this aspect of the problem, oil is *not* scarce and no one has benefited more from that fact than first Venezuela and then the Middle East. To boot, what is shaping the current situation, and has thus become the *leitmotiv* of this study, is that it is the oil with the lowest cost which is in most abundant supply.

To preserve their traditional margin OPEC would have to develop into a tight and fully-fledged price cartel whose members would have had to settle among themselves all inherent competitive tendencies so as to confront the rest of the world with a cast-iron common front. There is no need to dwell on the difficulties of finding a common denominator for the several interests – not only the international oil companies are animals very different from each other – it is still more unlikely that a militant association of producer governments would last for any length of time. Indeed if the oil producing countries were to take upon themselves altogether the responsibilities hitherto shouldered by their concessionaires they would be confronted with the facts of life, and would soon realize that *plus ça change, plus c'est la même chose*. In the first rush of enthusiasm, of feeling of achievement, they will all consider each other as equals (within the terms of the agreement of course). As time goes on they will see that – given low cost and, by definition, high prices – there is an enormous advantage to be gained by straying just a little from the straight and narrow path, in fact it will be only too tempting to try to be, as in Animal Farm, more equal than the others. Moreover, pressure calls forth opposing forces: a tight organization of producers will activate consumer interests and they could offer glittering prizes to anyone who would make concessions on price or terms simply by increasing their offtake from such country.

From that point onward it is only a question of time until the agreement turns out to become eroded to the same extent as was in its day the tacit conformism of the oil companies. This indeed is where we came in: potentially abundant supply remains the basic fact which, *if it stays unchecked*, will undo the attempt to earn a 'semi-rent' for any length of time: in such a situation to try to keep prices high, as compared with physical cost, is a labour of Sisyphus: the rock will topple down the slope every time one has pushed it (almost) to the summit.

To have said all this is not to deny the importance of an organization such as OPEC. Its recent Explanatory Notes to which I have referred show an attempt towards a flexibility of thought which has so far failed to express itself in the

Resolutions themselves. OPEC may yet provide a testing ground for new forms of co-operation, simply by raising the level of the negotiations between the parties concerned. The reference to 'a rational price structure to guide their long-term price policy' in one of their recent Resolutions may be a straw in the wind. I must repeat, however, that it will not take long to become evident that endeavours such as the one to correlate oil export prices in some way with those of 'goods which Member Countries need to import' will make sense only if they are an integral part of a more far-reaching system of planning. This, incidentally, would be equally true of the latter-day oil company concepts of a guarantee of revenue to producer countries. Even two swallows do not make a summer.

Conclusions

What then is a constructive approach to our problems? Before we could visualize it we had first to dispose of the thick undergrowth of prejudice which had been allowed to obscure the woods.

In the circumstances the greatest effort in shedding taboos will be required of the producing countries if only because their prejudices have become part of their pride. They will have to realize that there is no natural law which provides for the sale of a commodity in abundant supply at prices which are many times its cost and that they have simply been the beneficiary of the fact that at a given moment several interests, mainly those of the oil companies, their alleged adversaries, converged and resulted in a curious situation; and that the threat to the maintenance of this singular set of circumstances stems from the unrelenting pressure of straight economics. If they really want to become responsible for their fate – instead of maintaining their facile rôle of backseat drivers – they will see that they have to curb some of their desires to be able to fulfil others, indeed they will have to become part of the oil industry in a more fundamental sense than they ever had in mind.

The oil companies, chastened by their current difficulties, will perhaps draw the conclusion that, since they cannot organize their several interests in co-ordination with each other, the alternatives are either all-out competitive pressures, partly exerted by state-backed organizations of which Sojus-neftexport is only one, or increasing reliance on national and international authorities who are the only ones who could in the circumstances provide the framework of equitable planning.

As time goes on the British and French authorities are likely to abandon what has by now become an illusion, their self-assumed rôle as oil producers abroad, and will acknowledge their position as big oil consumers. By then the outlook of consumer countries will be clarified – some of them, if recent resolutions in Brussels and Luxembourg prove to be symptomatic, are about to liquidate the toughest taboo of all: the idea of safeguarding indigenous coal by making imported oil dear.

Should all this happen a number of problems would become manageable:

(1) A plan for the exploitation of oil reserves could be drawn up starting from an AS IS position to give historic positions their due weight; meeting increases in consumption, however, would to some extent be the domain of new suppliers, countries as well as companies. The recompense of such policy for the now paramount countries and companies would lie in their resources and organizations being safeguarded against extreme compression of profitability. A continued flow of risk capital into further exploration for oil and natural gas would be secured by a measure of certainty that new oil thus found would not remain altogether unsaleable or could be moved only under conditions of either uneconomic price-cutting or ruinous investment in downstream facilities.

(2) The diminished pressure of incremental crude oil production would tend to reduce the distortion of refining and marketing activities mentioned earlier on. It would again be possible to carry on such activities *per se* without recourse to producing-phase profits.

(3) Such stabilization of the producing end of the industry would be acceptable to consumer countries only:

(a) if the price level would not significantly exceed the present *actual* realizations;

(b) if consumer countries would have their say in the planning of oil production. In such circumstances, when producer countries and oil companies guarantee supplies at reasonable prices, consumer countries – with their industrial and financial resources – could in turn guarantee the producers a certain level of offtake and income and also supplies for *their* requirements on equivalent terms;

(c) if their nationals were to have fair opportunity to engage in exploration and production without undue discrimination.

(4) To establish such equilibrium there is no need to 'wait for Godot', for the advent of an international Authority which could set up a world-wide organization to run the show, or deal with the actual execution of a plan. The Canadian example shows that general directives, if only they do not run counter to the economic forces extant, can provide the framework into which the working parts are built by others directly involved in the day-to-day operations. Thus the international oil companies, whose forte is that they exist but who, if they did not exist, might have to be invented, and for whose technical organizational and financial performance there is no substitute in sight, would hang on to the main part of their functions, at least to the same extent to which they have retained them in the domestic sector of the US oil industry. All this is nothing new; the concluding sentence of my 'Essentials of Petroleum', written seventeen years ago, was: 'Whereas decisions of a *strategic* kind cannot be evolved but on the highest level, *tactical* decisions are best left to the industry itself'.

Some Common Carrier of Common Sense should become fit to initiate and guide such type of planning provided there develops a certain mutual goodwill generated by the preference of all concerned for standing together rather than for falling separately. In such a set-up groupings like OPEC, or the Committee which the Interexecutive of the European Economic Communities have in mind, would have their natural rôles to play, supplementing yet not supplanting the direct initiatives of the countries and companies involved. To achieve this we only have to stop fiddling and to get on with the job.

9   Topical Problems

This 'topical' comments on the terms of the Royalty Expensing negotiations between OPEC and the companies which were finally accepted by OPEC at a meeting in November 1964. What is not stated in this piece is that the 'compromise acceptable to all concerned' nearly broke OPEC. Iraq refused to ratify it and, indeed, their minister had authority to leave OPEC. Iran was, it is true, 'more co-operative' in the end, but only after the Shah had, under pressure from the companies, some of his own ministers and US/UK diplomatic representation, withdrawn his support at end-1963 from the agreed OPEC line and from the Secretary General, Rouhani, who had been instructed to carry it out. OPEC's defeat – for that is what it was – reflected the economic and political realities of 1963–4 and it was not until 1970, in very different circumstances, that they were able to gain the initiative.

During the last few months the negotiations between the oil companies and the producer-country governments have developed in a way which now makes it possible to review the general position of the two sides.

The last time we dealt with this subject in Topical Problems we expected that the divergency of opinion between OPEC and its member countries on the one side and the major international companies on the other would not lead to a unilateral action by the governments which would affect the existing system of oil concessions, but would in the end lead to a compromise acceptable to all concerned.

The reasons why our forecast proved correct are based firstly on the fact that the international oil companies are the only effective means for the oil-exporting countries to reach painlessly the free markets of the world, that is mainly Western Europe and Japan. In all these markets the oil companies are solidly entrenched either by their fully-integrated status (transport, refining and marketing) or by long-term agreements. As long as the governments of the main oil-importing countries do not take over the role of deciding where oil should come from and what price should be paid for it, it is the oil companies who retain what has been called the 'power of disposal'. Since nobody else but the companies can guarantee to a producer-country regular offtake and tax payments on a high level (say 80¢ per barrel), no country can afford to antagonise the companies too much. This is especially true as long as each country cannot be absolutely sure that one or the other OPEC member will not make a separate deal with oil companies – it is obvious that the oil companies could if necessary promise special advantages to any OPEC member who might be inclined to do so.

During the last few months Iran has been more co-operative in respect of the oil companies' proposal than some other countries, especially Iraq. The conciliatory tone of the latest OPEC resolutions is probably due to the influence of Iran, to which some of the other governments responded favourably, but also to the fact that the oil companies improved their offers, especially on expensing of royalties, sufficiently to make it impossible for the OPEC members to take up a militant attitude and to risk the solid advantages which the companies' offer provided.

Although OPEC is apparently going to study once more the pricing problems, it is not likely that its members will press any further the suggestion that posted prices should be brought back to the 1960 level, i.e. to that prior to the last reduction of the posted prices. On the other hand, there does not seem to be any tendency of the oil companies to pursue the idea of severing the connection between posted prices and tax liability. Consequently, it is unlikely that posted prices could go down in the immediately foreseeable future.

In this phase of the negotiations it looks, therefore, as if by an ad hoc compromise, which should provide additional income for the Middle East producer-countries of about $100 million in 1964 and up to $170 million in 1966, it has become possible for all concerned to delay any decision about fundamental changes in the relationship of producer countries, oil companies and consumer countries. To quote from an OPEC statement of about a year ago, the fact remains, however, that the real problems, when they are not faced or solved, remain; these problems are still derived from the fact that there is more actual and potential crude oil production capacity in the Middle East and North Africa than can be sold in today's circumstances. Consequently, competition between oil companies and, at a certain stage, between oil producing countries, will remain and, in the absence of agreements or regulations, will affect the investment, sales and taxation policies to be applied.

The first opportunity for these problems to come to the fore will be the decision on the extent to which potentially oil-bearing areas will be opened up by way of new concessions or by direct activities of state-owned companies. The choice of concessionnaires and/or partners will indicate whether the tendency towards more intense competition will make itself felt slowly and gradually or whether there might be in only a few years new groups which will have to develop reserves, should they find them, rapidly and without regard to the present balance of power among companies and countries.

10 Public Enterprise

In 1966, Frankel published his study of the Italian oilman and chairman of ENI, whom he had known personally and for whom he had a high regard. Obviously, long extracts covering the life of Mattei or the history of ENI are inappropriate to this volume. However, the concluding chapter of the book is a general commentary on the nature of public enterprise which well expresses much of Paul Frankel's own philosophy and exhibits his interest in and sympathy for the lessons of history, politics and sociology. Frankel's subject and expertise has always been the oil industry, but an oil industry that belongs to international politics, economic realities and historical processes. He has, more than most people connected with the industry, been its philosopher as well as its critic and commentator. This chapter (here in an edited version of the original) stands in its own right as an expression of some of his thinking, echoes of which will be heard in other pieces in this volume from both earlier and later writings.

The 'Mattei experience' gives rise to the fundamental question: what is the *raison d'être* of public enterprise in non-Communist countries and outside the public utility field? Does the government, to make its weight felt, have to enter the ring and become a combatant rather than remain the referee?

There is the extreme liberal conception of the state's staying aloof to the extent of leaving trade and industry virtually to their own devices which are determined by what is called the 'market place' in which rules the law of supply and demand; thus governmental action or reaction is to be confined to looking after public safety and national security. Others visualise that the authorities have to go further and should influence economics by setting deliberately the rules of the game, i.e. by their fiscal and custom tariff policies and by their

Part Three of *Mattei: Oil and Power Politics*, Frederick A. Praeger, New York, 1966.

influence in the money market. Such measures can be applied in different degrees of intensity, ranging from the liberal to the protectionist. But however illiberal a country may be, for instance by virtue of a high-tariff customs policy or of quantitative controls, it is essential according to this school of thought that governmental measures are of a *general* nature applying to whole categories of trades and traders; in fact, they provide only the basic framework relevant for anyone to whom its terms and definitions apply.

For the government to take *specific* action by protecting or vice versa by penalising individual enterprises or, going further still, by becoming judge in its own case, as it were, by turning itself into a competitor of its subjects, means a fundamental change of scope and status.

A Few Test Cases

In our era and in non-Socialist countries, the state enters business mainly under conditions of stress; such decisions are taken either in the course of war or under political siege conditions. Otherwise, it is a case of private enterprise gone wrong, asking to be bailed out by the government: hence the old saying, that the capitalists are always in favour of the nationalisation of their losses.

Such was the origin of the wholesale takeover of banks and industries in Italy between the two wars which led to their being consolidated in the *Istituto per la Ricostruzione Industriale* known as IRI. There had been a virtual breakdown of the spirit of private enterprise when its predestined leaders, the big banks, were found wanting at the time of the great depression. In Italy (as, incidentally, in Austria) no one but the state was at the time able to fill the void which threatened to suck in the whole economic structure.

At the outset it was assumed that this was but a holding action and that the return to 'normal conditions' would be marked by the 'reprivatisation' of those enterprises which the state had taken under its wing. One could say that the interventionist, or, to use the nowadays more fashionable term, *dirigiste* style of the Fascist regime tended to frustrate such a trend of events and that the Second World War with its aftermath held it up once again. There is more to it than that, however, and the continued existence and prosperity of IRI and many of its operating companies are due to other than straight political reasons. In IRI there has been developing a curious compromise between public enterprise and private investment interests which has some general relevance, although it probably owes much to the specifically Italian genius for improvisation which is at its best when it is not weighed down by dogmatic considerations.

One of the main reasons why the vast industrial complex of IRI was not returned to private ownership was that there was no Italian capital market which could have taken up the bulk of the IRI assets as they were at the time – it was obviously undesirable from the state-holding point of view to let the assets go to foreign interests or to sell them off piecemeal, which would have led to the early disposal of easily marketable items, leaving behind a hard core of problem cases.

It was not Socialist influence – such as it was at the relevant times – which kept the IRI companies in the state's orbit; for very good reasons IRI, or rather what it stood for, never became a tenet of Socialist intentions. It was generally recognised that if the Italian heavy industry which formed the centre-piece of IRI was to be in private hands, it would either gravitate towards one or another of the existing agglomerations of industrial strength and entrepreneurial ability – mainly Edison, Montecatini and Fiat – or become the nucleus of yet another strongpoint of political influence.

This is perhaps the appropriate moment to talk of the nationalisation of electric power generation in 1963 which, leading to the formation of ENEL (*Ente Nazionale per l'Energia Elettrica*), was one of the preconditions of the left-centre coalition formed at the time. No one seriously claimed that there was an imperative need to make power generation more efficient by a degree of centralisation of management or investment policies such as only straight nationalisation could provide; Edison and a few other companies involved were highly effective and their co-ordination could hardly have been bettered. It was a fact, however, that they had through the decades occupied a position of *political* power, a fact which now acquired a boomerang quality since it made them particularly obnoxious to some or most of the parties whose support was necessary if a left-of-centre government was to be formed – and there was no alternative combination of parties in sight at the time. On the other hand when ENEL was to be set up there was an outcry even in circles not averse to its formation: 'Not another ENI!'; that is to say, ENEL was to be neutralised and not to become one more point from which politico-economic influence could be wielded.

One could have countered this apprehension by saying that power generation and distribution, being a classical public utility, fell into another and less controversial category anyway. This distinction between the public utility and the competitive sectors, however, is now less appropriate than it used to be. No one has today the monopoly position the railways or coal mining each used to enjoy in its heyday: rail now competes with road and with water-borne transport, and latterly with pipelines; even a monopoly of electricity generation has to compete with oil and gas. Hence the lively (and sometimes noisily conducted) rivalry of the coal, gas and electricity industries in Britain – all nationalised – and, after all, Mattei's natural gas could also have been considered to be in the sphere of public utilities.

To revert to IRI: it has managed to remain detached from the immediate political sphere. At its best it combines some positive factors of both private and public enterprise, and only at times and in certain difficult instances has it made the worst of both worlds. Its management is 'technocratic' in a functional sense – a state of affairs which was probably helped to emerge by the decision to have (or the need for) private investors taking up minority participations in IRI enterprises. For this and other reasons IRI was and still is run as if it were a classical enterprise which has, however, substantial backing of the state in return for which it orientates its investment and managerial policies towards certain general principles in line with the trend of national affairs.

IRI is essentially a case of public enterprise filling the gap due to a breakdown of the working of the private-enterprise system, but it must be admitted that had the government not intervened or had it tried harder to divest itself of the assets it had taken over, since nature abhors a vacuum, some sort of private enterprise *would* have emerged in due course. Its character or performance might have been undesirable but it remains true that, by taking over, the government has inhibited the formation of a kind of new equilibrium. It is also true that the existence of a significant public sector in the competitive field tends to modify the character of private enterprise operating therein, not always in the direction of greater efficiency.

All this is probably true also in the currently relevant case of the developing countries where the emergent governments tend to occupy the Commanding Heights of the economy, claiming that there is no, or not enough, home-grown capital and entrepreneurial talent available: key industries involve a great deal of long-term investment with moderate returns, whereas indigenous private business circles are used to quick turnover and high returns on their capital. Direct foreign investment (if available) raises the question of national prestige and of political power, hence the preference for channelling it through state-controlled enterprise.

There are, however, some developed and otherwise private-enterprise-minded countries in which public enterprise has taken root in the course of the last decades. In Austria there were several waves of governmental involvement in business: there was first the one resulting from the great depression of the thirties, which started, as it were, with the fall of the Vienna Credit Anstalt, the focal point of Austria's economy, which was taken up by the state and has remained in its orbit ever since; after the *Anschluss* came the war-time economy and industries were put under control of the *Reich* or were started from scratch by it, like the Hermann Goering-Werke. After the collapse and the four-power occupation, the assets previously belonging to the *Reich* or to German enterprises became a kind of no-man's land and were vested in the Austrian state; the diverse reasons for these enterprises having been taken over at different times resulted in their being still more of a mixed bag than was and still is IRI, and the peculiar Austrian set-up has tended to frustrate any attempt to provide a rational system of planning and management.

As Austria was ruled from 1945 to 1966 by a coalition of two almost equally strong parties, one of which professed belief in capitalism, the other in socialism, the *status quo* was (uneasily) maintained by a kind of 'standstill' agreement: the national-ised part of the economy was kept in being without further encroachment upon the private sector being permitted.

In spite of the general prosperity of the Austrian economy which was achieved once the aftermath of the last war had passed, the overall performance of the nationalised sector does not appear to have been outstanding. This was due partly to the inclusion of some genuinely and irretrievably shaky enter-prises, but also to the peculiar system which post-war Austria devised in order to cope with the fact that its two almost equally strong parties, who did not – as they do in Britain – form single-party governments in turn, were locked for a very long time in a form of coalition. This way of life has un-doubtedly considerable merits inasmuch as it has eliminated the dangers of virulent party feuds which marred Austria's inter-war years. In the sphere of public enterprise Austria has also saved itself the near-comedy of steel nationalisation and de-nationalisation which Labour and Conservative regimes in Britain have played as they succeeded each other. Yet, the party contest had cropped up in the next layer and led to the peculiar Austrian pattern of representation of the two parties in the management of public enterprises proportionally to their strengths – hence the vernacular term: *Proporz*-System.

To carry the checks-and-balances system to such length does not make for vigorous and go-ahead leadership, but the fact that if the Number One of an enterprise belonged to one party his Number Two had to come from the other did limit the opportunities for outright corruption; on the other hand, it carried political 'wheeling and dealing' right down to day-to-day management. What made that *modus procedendi* still more questionable was the fact that the machinery of the country's administration itself had become imbued by the all-pervading influence of the political parties: civil servants were also quite often appointed 'proportionally' and were thus no longer altogether beholden to the nation as a whole – or to the state, whatever that may mean – but to their respective party which had made and which could undo them. *Quis custodiet ipsos custodes?*

The saving grace of this system was the innate Austrian

flair for making the best of any situation by not taking it too seriously and for getting on with the job in hand. Some enterprises were weak and lifeless, they were carried on mainly for political reasons – but that is not a specifically Austrian phenomenon. Furthermore the lack of buoyancy of most of them eliminated the danger of their managers acquiring political power on the strength of their jobs – there seem anyway to have been a singular lack of strong personalities: if there were any around the prevailing climate did not favour their coming to the fore. It has been said that public enterprise in Austria could do with at least some pocket-size Matteis. At the other end of the performance scale, the Austrian system provided some examples of first-rate management and of technical innovation – *vide* the revolutionary 'LD' process of steel-making which has now been adopted all over the world.

The drawbacks of such a system are obvious but it is worth considering whether the situation in some other countries, where the government encourages new enterprises and safeguards existing ones by taking the bulk of the risk upon itself (thus guaranteeing to private enterprise a certain return on its capital, whereas profits made over and above that minimum belong mainly to the entrepreneur) is really more equitable and more workmanlike. There the government takes all the risk of failure without prospects of benefiting from success.

The origin of public enterprise in Germany can be traced back to 'cameralistic' influences left over from the pre-capitalist era – hence the existence of some long-standing state enterprises in the mining and other sectors. Yet right up to Hitler's days state-owned industry was no point from which power was wielded – it was the other way round: policy was determined by the entrepreneurs, the Ruhr barons and the wizards of chemistry.

In Professor Erhard's post-war Germany the remnants of state enterprise were considered as a somewhat embarrassing leftover from a disreputable past and were kept in the background by being run as firms like any other and, with a few notable exceptions, with rather below-average pressure; their managers were expected to look after their balance sheets and nothing else, on the other hand they were not subjected to the discipline imposed by a multitude of demanding shareholders.

As soon as possible these enterprises were expected to be 'handed back' to private ownership anyway. Alternatively shares were to be offered to small investors which, in the nature of things, had the character of debentures; the need to service them made the management of these enterprises still more independent of their main shareholder and any fancy ideas the latter may have had in mind. It is an intriguing thought that even in this uninspiring and antiseptic atmosphere leadership of a man like Professor Nordhoff could result in as remarkable a success story as that of the Volkswagen, surpassing even the great achievement of M. Pierre Dreyfus, another outstanding industrialist, at the French *Régie Renault*, the nationalised motor car makers.

It is in France where public enterprise has taken on a particular significance, inasmuch as there neither the failings of industries and the weakness of the capital market nor the lack of an entrepreneurial class forced the authorities to step into the breach. The case of Renault, where the state became unwittingly heir to the late owners' estate, is an exception: generally speaking, French public ownership of industry fanned out from its traditional centre of public utilities and transport into the adjacent spheres of energy supplies at large.

The origin of the French tendency towards public enterprises (such as it was and such as it is now in somewhat different circumstances) never did stem from socialist leanings in the party-political sense. It is rather the result of a general conception of the state's duties and prerogatives: such conceptions anchored in a truly permanent civil service which, irrespective of the continually changing and shifting party assortments and governments of the Third and of the Fourth Republic, maintained the continuity of the French state and nation. Thus French public enterprise was initiated by civil servants and run by a curious, but in many cases remarkable, brand of business administrators dedicated to their task, efficient, intelligent, single-minded and incorruptible but, perhaps for the same reasons, opinionated and arrogant. That French public enterprises attracted a certain *élite* of industrialists might to some extent have been due to the scope for them there to fulfil certain technocratic ambitions for which private investors were not (or not yet) ready. There is, however, another essential element: the opportunity of doing a big job, which the man of action cannot easily resist, was enhanced by the underlying knowledge that the effort was expended not to swell the coffers of some capitalist or other, but directly in the interests of the nation – 'All this and Heaven too. . . .' This phenomenon is not confined to France: in Britain nationalised coalmining, railways and power generation have drawn into their orbit a galaxy of men such as Beeching, Robens, Hinton and Edwards who, coming from very different ways of life, had in common the idea that they were not there merely to administer what there was, but to set a high standard of industrial performance as a service to the community.

Anti-Trust

A high-ranking Frenchman once told me that the peculiar system of their public enterprise, especially in the oil sector, was in fact nothing less than a European version of the American Anti-Trust philosophy; by coining that phrase he provided also a highly relevant description of one of the aspects of Matteism. Seen from this angle the direct intervention of the state is dictated by the need as it is seen by its advocates to provide countervailing elements against agglomeration of economic and political power in the hands of large-scale private enterprise. Such concentration of trade and industry in big units is held to be bound up inevitably with the advanced stage of industrial development, and if a proper balance is to be maintained the authorities have to counteract (or to neutralise) such impact by appropriate means. What is considered appropriate depends entirely on the historical, social and legal background against which any measure taken has to be seen.

The really potent argument against 'bigness' in business, as well as in other spheres, does not stem, or at least does not stem mainly, from utilitarian considerations, nor does it involve an adverse verdict upon its effectiveness. The aversion against bigness has deeper origins; it derives its strength from the instinctive desire to redress the balance between otherwise unequal forces.

The way such conceptions manifest themselves varies according to the prevailing political climate in a country, and according to the structure of its industry. The Anti-Trust policies in the United States do indeed provide a telling example: in an enormous area of free trade there was an unprecedented opportunity for large-scale enterprises, which, once they had attained a certain position, tended to grow bigger as they went along; if there was, to boot, a chance for a few paramount units to co-ordinate their plans, they might achieve preponderance to the exclusion of others, a state of affairs which could lead to two equally unpalatable results: having reached a state of near-monopoly, they could, to some extent, set their own price levels for their goods and services and, more important still, their financial strength and the comprehensive sweep of their enterprises might provide opportunities for attaining direct power within the body politic of the country itself.

That in the U.S.A. the defence against such twin menace was organised by way of judicial and not of administrative procedure is a result of its constitutional set-up. There exists a vast literature on the sociological, the economic and the legal aspects of Anti-Trust principles and this is not the place for even a brief analysis of all the problems involved, but some of their features are of direct relevance to our main theme.

Whereas it is the pretension of the concept of free enterprise that it is self-regulating, and therefore, in theory at least, is neither in need nor does it allow of outside interference, the idea of far-reaching remedial action, which is the mainspring of Anti-Trust policies, does involve the admission of the need for measures which do not originate from the profit-orientated motives of the operators themselves. It cannot be said that Anti-Trust regulations are designed merely to curb abuses and are thus almost on a par with safety regulations or with sanitary precepts – they are no less than a deliberate attempt to impede tendencies towards bigness, concentration and market control which would prevail to a great extent in the absence of countermeasures, which do not belong to the sphere of economics as such. There can be little doubt that, at least in the first instance, such curbing of business-orientated trends towards concentration involves a sacrifice but that, at least in the eyes of the advocates of such policy, it was worth the price – only recently it was said in the U.S. that 'The

Supreme Court has expressly recognised that the fragmented economy is desirable . . . even though inefficiency may result.'

The trend towards concentration and towards co-operation of competitors must be a true ground-swell to render necessary such an elaborate and recurrent (if never altogether successful) treatment of an orthopaedic or even of a surgical nature. Whoever has watched the way the Sherman Act and the related U.S. laws have been interpreted and enforced knows that the Attorney General does not usually stop at seeing that certain general rules are being observed, indeed the Department of Justice tends to become very specific: it does not only object to certain actions, it actually requests companies to take definite steps, which it maps out for them in great detail. That all this is done in a way which can be enforced in the courts is characteristic of the U.S. historical and constitutional set-up, but the case law has in this respect been modified so much as time went on, and in the last decades so rapidly, that the 'social engineering' character of the Anti-Trust doctrines has not had time to become properly neutralised. Anyhow, in other countries and in different circumstances the same or similar policies may be, or may have to be, pursued by different means.

Now we can see more clearly what my French friend had in mind: the concept of state-owned enterprises, or of companies backed and to some extent controlled by the authorities, owes some of its rationale to the idea which is also inherent in the American Anti-Trust principles which involve state action to safeguard competition in the private enterprise sphere. Economics being what they are, it is any government's role deliberately to establish a situation in which the various operators can, as the saying goes, compete with each other *on equitable terms*; the trust busters have in this respect much in common with state enterprisers. The door which leads to true competitive freedom seems to have to be secured by iron-clad warriors. Hence the dictum of Lacordaire, a French religious thinker of the mid-nineteenth century: 'Between the rich and the poor, between master and servant, it is freedom which enslaves and it is the law which liberates': a dictum used by an Algerian Minister of Trade to crown his political argument at a 1964 U.N. Conference.

Remarkable is the constancy of French political thinking and experience; more than a century after Lacordaire, in a debate on the Vth Plan, M. Pompidou, de Gaulle's Prime Minister, used language that has to be read in its original to be savoured fully: '*La loi du laisser-faire . . . sœur de la fatalité, alliée de la richesse et complice de l'injustice. . . .*'

Mutual Interference

Before drawing conclusions from what has been set out so far, we have to place it in the proper context:

We do become quite often prisoners of our way of speaking – by talking of 'governmental interference in economic affairs' we have prejudged the case. The term 'interference' implies that there exists a self-contained sphere of economics – a sphere in which economic factors lead to certain well-defined developments and where the government intrudes from the outside.

In fact, there is no separate realm of economics, just as there never existed the *homo œconomicus* – a creature motivated only by its straight monetary interests – on which the writers of the late eighteenth century classical school of economics depended for their claim that Laws of Economics could be discerned which would give their models a truly scientific character. The idea of a self-contained sphere of economics is a useful working hypothesis but nothing more: it is obvious that productivity as we know it would be greatest if all elements of the production process could be applied in a way altogether orientated towards the maximum results measured in terms of quantity or of money value. This system, foolproof as it is within its own compass, presupposes, however, that no other considerations are allowed to make themselves felt, i.e. that all elements involved – including the human ones – are interchangeable and expendable. Thus people would have to be free to go and live wherever their work would be most productive, indeed, there would be no alternative to their doing just that; also, people would have to accept wages as determined by current supply and demand, and put up with unemployment as long as their labour was not wanted at the time.

Yet, in our own time at least, the human element is not to be considered mainly as a means of production; in fact, people consider themselves in the first instance as consumers, and in most countries they *are* electors. Therefore, what we have is a combination of policy decisions, a blend in which straight economic and human (often called 'political') considerations play their respective parts. It is therefore misleading to consider the economic sphere as a centrepiece, whereas all other elements find themselves relegated to the periphery – their being brought into play is considered as 'interference' with a, or, worse still, with *the* natural course of events. One could be equally justified (or rather, one would be equally wrong) in considering what we call social or national considerations to be the 'real' ones, with economic elements intruding into their territory. The truism that our life consists of a combination of all these elements has to be emphasised here only to show that the vocabulary we use has been formed by a one-sided way of thinking. To see the several considerations which together, by way of co-existence, form the relevant pattern is to realise that neither the species we call political, nor the one labelled economic, can be developed too far without regard to the other: if we did so, we would outrun our actual means in the first instance, and would alienate the people altogether in the second.

The French view, mentioned earlier on, was that general rules, such as the laws against cartels and restrictive practices, passed in some European countries during the last twenty years, were not adequate, since they did not and could not have the 'teeth' built into the U.S. Anti-Trust system. The thesis that positive and specific governmental action was required hinges on the argument that the government could influence the development of industry by way of investment and that of trade through a pricing policy only if it controlled directly a significant sector of the country's economic life. There is a subsidiary argument which runs that the authorities cannot acquire inside knowledge of the working of industry if they are not represented there by *entreprises témoins*, test cases, as it were, whose business activities (and one supposes whose motivations) the government could survey like an open book.

Pulls in Either Direction

This thought leads us back to the main question: what is the object of the whole exercise, what is the job of public enterprises in those spheres where they are designed to compete or to co-exist with private enterprise? Are they to be orientated solely towards their own economic success as it can be measured by their Profit and Loss Accounts, or should they become direct instruments of the social and trade policies of the government which has set them up and which has the right to control their planning and operations?

Actually Mattei was always faced with the problem of what his enterprises were meant to achieve, and in ENI there were all along two schools, the one putting the accent on the *public* character of ENI, emphasising its role as the strong arm of the state, whereas the other wanted to look after the *enterprise* first, maintaining that if it was sound in itself it would also be a more effective weapon if and when needed. Mattei himself never quite came out on either side of the great divide; here again it was for him *l'uno e l'altro*.

In fact the direction in which the public enterprise faces is relevant for the way things will go altogether; a strong tendency towards stressing the state-determined policy of the enterprises will lead to more and more of the same medicine being prescribed and administered in the economy at large. In order to render viable undertakings which derive their code of conduct from 'political' considerations, their sphere has to be substantial and self-contained, also an economic climate has to be established into which they can be fitted successfully.

Any marked tendency in that direction will progressively lead to the establishment of a centrally planned economy in which the planners have powers to direct and in which the non-public enterprise will tend to take on an auxiliary character.

If, however, public enterprises are run with the idea of having an economic existence of their own, the prevailing winds will drive them more or less rapidly towards a character and style which will make them lose their peculiar identities by merging into the private-enterprise landscape which surrounds them.

There can be little doubt that post-Mattei ENI, whilst it continued to have the backing of the Italian State and worked along the lines of the Italian national interest, had, in order to survive the predicament in which it found itself after its founder's death, to take the route towards an economically balanced enterprise *per se*.

In this, as in other respects, we are somewhat assisted by the knowledge of the scope of the policies of companies in the oil industry sphere that were sponsored by their governments, and of these policies' shape, of their successes and failures: rather than add to the vast literature on the general aspects of public enterprise I propose to take another look at their record to date in Britain and France.

The British, who tempered their bold essay in state enterprise in a highly complex industry – the acquisition of Anglo-Iranian in 1914 – with pragmatism and common sense, as we saw earlier on, left their bastard to fend for itself after having set it up in business on a lavish scale. The board members nominated by the government have never been known to interfere in the business of the company – although there were some rumblings of mutual discontent when the Labour Cabinet under Wilson came to power. The experience of what is now British Petroleum has, however, only limited general relevance: firstly because the innate discipline of the British made it possible for the government to rely on the ordinary directors of their company to take overall national interests as fully into account as they knew how; secondly – and this is infinitely more relevant – that company, due to its endowment but also to the continuing and outstanding buoyancy of the oil industry, never needed any actual help, to which strings would have been attached, nor did it ever need government-borne finance of any kind. The company generated (or on rarer occasions attracted) all the funds it could possibly need, and thus the government as a shareholder was, at least until 1966, excused from taking part in the fundamental decisions, i.e. the planning of investments.

The foremost French test case – *Compagnie Française des Pétroles* – has something in common with that of its British counterpart; its endowment with the Iraq Petroleum Co. share provided, as time went on, the opportunity for large-scale self-financing. In other respects the French experience was different from the British one: whereas B.P. was one competitor amongst others, the privileged position in which C.F.P. was put in the domestic French set-up gave the company a chance to build up an organisation which made it less dependent on sustained governmental backing than it might have been in its early years. On the other hand, France's oil regime provides some proof for my statement just made, that deliberate development of public enterprise, with the accent on 'public', tends to spread in one form or the other: because C.F.P.'s position had to be buttressed, all other companies too had to work within a system of safeguards which made it exceedingly *dirigiste* all round, but also remarkably remunerative for the companies operating within it. When some spokesmen of private oil companies complained that they were living in a cage the reply was that at least it was a gilded one.

The case of the French oil system as a whole provides some other very relevant indications: faced with the choice to sustain home-grown enterprise by general protective measures or to establish state-controlled, i.e., in the French parlance, 'national', enterprises, the authorities adopted the two methods rolled into one. In a general system of regulations which, as I have just said, determined the status of all concerned, there was, according to policies which changed surprisingly little from the Third to the Fifth Republic, still the need for a number of enterprises directly beholden to the French *Pouvoirs Publics*. Thus the French embraced the whole range of measures from the permissive through the protective right to the positive type. The fundamental question of sustained supply of capital came to the fore only when, after the last war, the French government wanted to broaden its approach and embarked on an ambitious programme of exploration and development in Metropolitan France and in the French Franc Zone, mainly in North Africa.

Although Saharan oil later on did attract some considerable private investment – which many of those who made it soon had reason to regret – there would hardly have been sufficient promise in the first instance for it to get the exploration going at the speed and on the scale which the French government wanted to see, mainly for political reasons; that was the period when one thought that Algeria could be made safe for France by developing massively and rapidly the potentialities of the Sahara. As it so happens the long-odds chance of oil and gas in the Sahara came off, and at least some of the funds put into it in the early days provided satisfactory returns in the straight business sense of the word. Although subsequent political events might yet compel us to qualify such assessment, the experience of B.R.P. (*Bureau de Recherches de Pétrole*) and of the other state-owned French oil enterprises provides an example of successful endeavours in which the authorites had to take the initial decision and bear the specific risks involved in an enterprise in its fledgling days, but could progressively reduce their backing and direct control once some of these enterprises came into their own. Here we have something of a rebirth of the nineteenth century conception which gained currency first in Germany, of the need for an *Erziehungszoll*, a temporary customs duty protection for emerging industries, required in countries which had to face then overwhelming competition from those who had gone first through the experience of the Industrial Revolution – mainly England. Although industries thus encouraged and cosseted were only rarely ready to de-clare themselves grown up and no longer in need of care and protection, there were cases of complete viability once their teething troubles were over. None of these case histories in-volved the need to face in a fundamental way the crucial question: what does the state want to derive from becoming an entrepreneur – does it want to have effectively run companies which are to give a satisfactory return on the capital involved, or are they brought into and maintained in being to further national interests even if they are not profitable in terms of their own balance sheet?

Obviously if immediate profitability is the *only* yardstick the need for the state to get into the mêlée, except as a strictly temporary measure, is greatly attenuated – apart from the anti-bigness impact, why should it go to that length if all it means to do is to replace private enterprise in the sphere where it is at its best? On the other hand is the community to be the champion only of lost causes and do all the things which are 'uneconomic'?

If these questions were to be answered as they stand, there

would be little public enterprise beyond the public utility sector, but the real problem is more complex than that.

Managerial Tendencies

Apart from the cases where (*vide* the emergent Anglo-Iranian), an initial risk has to be taken for non-commercial reasons, there are the borderline cases between political and economic interests. Generally speaking, the managers of such enterprises will have the tendency to make them stand on their own feet as quickly as they possibly can; the less they depend on support out of the public purse – the idea of being subsidised out of 'the taxpayers' money' is a distasteful one – the more independent and powerful they become. The feudal barons were strong as long as they were sustained by their own local wealth and not dependent on favours bestowed upon them by the king.

No one knew that better than Mattei. The sources on which he drew to finance the rapid and sometimes indiscriminate extension of ENI did not include any substantial direct government grants – what was the equivalent of the share capital of a company, the *fondo di dotazione*, was and remained during Mattei's lifetime pitifully small; the cash came from the substantial profits on natural gas, from bond issues and later on mainly from bank loans – in each case the cash could not have been obtained without the co-operation and backing of the state but never could it be said that governmental monies had been directly paid out to ENI on any substantial scale. This contrivance had great advantages in the eyes of the public and, conveniently enough, helped to bypass the classical budgetary controls over the flow of funds towards Mattei's enterprises.

Commercial success of a public enterprise is pleasing all round, but the tendency to strive for it first and foremost may invalidate the intentions which led to its establishment in the first instance. The cases of B.P. and in a different way of C.F.P. show that by and by such management feels that it has more in common with private enterprise by which it is surrounded than with the authorities by which it has been called forth. Far from representing public interest in the field of economics, they become, willy-nilly, spokesmen and lobbyists of industry *vis-à-vis* the government: ironically they are then *entreprises témoins* in a sense quite different from, indeed contrary to, the original conception.

At the other end of the scale, a situation where a public enterprise is allowed for an indefinite period to make losses or to fail to earn an adequate return on the capital invested is also fraught with danger. Then it loses the benefit of the wholesome discipline imposed by necessity to provide for what one calls a 'payout': if the manager (and yet more the planner) is not responsible for the economic and financial consequences of his actions (or of his failure to act), if he can rely on the state to foot the bill, however high it may be and wherever it stems from, if a situation is allowed to persist which 'dulls the edge of husbandry', the enterprise is likely to sail like a ship without a compass and its leaders to lose all sense of proportion – social motives or the national interest are conceptions too general to provide sufficient guidance to actual management.

There is another side to all this, however. The government, as the trustee of the community, can deliberately embark on activities, knowing full well that in the short or even in the longer run they will not be justified by the financial return they offer; there will have to be cogent reasons for it – but is such line of action really so different from the policy of a large-scale and diversified group of companies which deliberately carries some activities or some affiliates which lose money because their existence provides advantages – tangible or intangible – to their sister-companies or to the group as a whole? This means in our terms simply that in certain types of business a given enterprise need not be the really relevant unit of operation and that it is the sum total of activities which matters. To make sense this thought presupposes, however, that one loses neither cognisance of the cost of a particular line of conduct on the one side nor, on the other, of its proper place on the higher level of decision making.

If it is the inherent peril of public enterprise that it neglects good housekeeping for the sake of long-range goals and motives, private enterprise, responsible to its shareholders, is faced with the opposite kind of problem: its management has to weigh up the immediate need of providing dividends against the task of going in for investments which can yield returns (if any) only at a much later date. Although many shareholders at times fancy 'growth stocks', even then they favour mostly shares of companies who also provide them with bread-and-butter returns there and then.

By observing the climate in which public enterprise develops, we have gone a long way towards defining the place which Mattei's enterprise takes up in the context of Italy on the one side and of the oil industry on the other.

National Interest – or Ambitions

However much his antagonism against the Italian private enterprise big business and the urge to provide and to make effective some countervailing power were the prime movers for Mattei, I have shown, I trust conclusively, that it was his relationship to foreign industrial, and thus to some extent political, power which made Matteism what it was for a time.

If the need, or at least the desire, to curb the extreme concentration of economic positions and thus of domestic political power calls for remedial action of one sort or the other, the justification of such action is all the greater where the pressure or threat of such concentration is of foreign origin.

Once again the question arises how far general measures such as protection of indigenous, and/or discrimination against foreign, enterprises can be adequate. Apart from the politically and psychologically invidious character such measures may have, there is some likelihood of their being in the end less effective than their advocates maintain they are; this is due to their essentially negative character – you can curb the incidence of foreign investment by a system of tariff and other barriers which impede the inflow of foreign merchandise and of capital, and you can establish ordinances limiting shareholding and management by others than your own nationals. Under the shelter of such measures indigenous enterprises may develop and prosper but, due to the lack of suitable entrepreneurs, this may not always be the case. Furthermore the 'closed-shop', or, if you prefer, the hot-house,

atmosphere which such necessarily far-reaching but rigid policies engender may affect altogether the degree of economic
efficiency and virility.

It is in these contingencies that there might be a genuine
call for governmental action rather than regulation, for setting
up and directing a national enterprise which, without exclud-
ing foreign or domestic private entrepreneurs, does carry on
its business in a way which is in the first instance determined
by motives closely in line with the requirements of the country
as they are seen at the time by its responsible authorities.

The call for such enterprise of national interest becomes
more pressing if one remembers that the eight big oil companies, although genuinely aware of their international responsibilities, are after all directly responsible to their own
governments when it comes to it. It is entirely understandable
that Rome, Berne or Vienna do not want to leave the handling
of one of the vital industries altogether to companies ultimately linked to Washington, London or Paris.

The pros and cons of ENI as a national enterprise living
alongside, and to some extent in opposition to, networks of
international enterprises based on a few (as seen from Italy)
foreign countries have been outlined. Now we should see this
problem against a more general background: the question
arises whether the formation of local – and in the initial period
inevitably small-scale – enterprise was a retrograde, i.e. anti-
economic step as measured by the operations of the comprehensive networks of the international oil companies.

'Should the Romans have stayed on in Ancient Britain?'
This question was asked and answered recently during a
public debate by Mr. Iain Macleod, that Philippe Egalité of
the British Empire. His backward glance was prompted by
the statements of other participants in the debate who had
pointed out that the newly-independent nations in Asia and
Africa would have been much better off had they remained
under their respective colonial regimes or in their association
with Western powers. Independence involved at best a more
primitive, more limited and altogether less effective administrative and economic machinery; it could even mean a relapse
into tribal warfare (disguised at times as a party-political
contest).

Whereas Mr. Macleod accepted such statements of facts, he reminded those who made them that the end of Roman military and administrative presence in Britain and the contingent fade-out of Latin civilisation had undoubtedly had at the time very grievous consequences and opened the door to a 'barbarian' way of life. Although it then took centuries to create a fabric of civilisation which could measure up to what Rome had provided for her outlying dependencies – did the other speakers, Mr. Macleod asked, actually regret the cutting loose from alien domination, the process without which the specific nature of the English would not have had the chance to take on its own shape?

There can be little doubt that the British and French overseas possessions, and, for that matter, those of Holland and Belgium, were more efficiently organised than are, or could be, the large number of 'independent' states which sprang up after 1945 in the territories previously under imperial rule or management. With hindsight we can now also discern that the peoples which became countries after 1918 in the process of the disintegration of the Austro-Hungarian Empire – which its apologists used to say would have had to be invented had it not existed – were forming economic and political units which could not be successfully sustained on their own for any length of time. In fact, after a brief interlude of independence, they were first forced into the Greater German Reich, only to fall into the Soviet orbit after the demise of the former. Some of the independent Asian and African countries too are now on the verge of becoming pawns in the big-power game. Yet Mr. Macleod was right: the self-determination of the minor nations in Europe and the quest for an identity of their own of the erstwhile subjects of colonial regimes were inevitable processes not to be measured only in terms of administrative and technical efficiency or even of current prosperity; there is no history without tears.

Here lies the main difference between Mattei's ENI and the IRI-type enterprises. It was not for him merely to manage an inheritance or to fill a vacuum adequately, he set out to fight to obtain independence first and then entry into what he saw as a closely guarded realm; to do that an extraordinary effort was needed and unorthodox means had to be brought into play. Having entered the scene by way of a political status of sorts he had to go on in the same vein and in the process he acquired political power which went beyond his immediate sphere. Such influence as he had was not obtained without cost; to get others to meet his requests he had to deliver the goods left, right and centre. Some of the more obvious ways and means have been touched upon, but there were others more onerous in terms of industrial logic; ailing engineering and textile firms had to be rescued at high cost and investment in plant had to be made because of the need to keep sweet some local pressure groups.

This two-way power play is at the centre of the problem: we have seen that governmental policies (or mere politics) can create and maintain conditions which deviate widely from what they would be if they were to be determined only by so-called straight economic factors. On the other hand those who manage the enterprises which have become carriers of such policies are in a position to acquire power; when dealing with 'Italian Politics' we have seen that in Italy there were no checks and balances by virtue of which one could have coped with the impact of power concentrations. This was forcefully brought out by Indro Montanelli in his articles in the *Corriere della Sera* which appeared only a few months before Mattei's death and to which Mattei failed to give a convincing reply. Whoever has had to wade through the morass of irrelevant press attack and apologia which the Mattei phenomenon has brought forth will appreciate that there at last was an analysis of some of the real problems involved. The author maintained that if a Mattei had been able to arrogate to himself a position which was not his to take up, if the servant of the state had made himself its master, the fault was in the system itself rather than with the man who, inevitably, by virtue of his vision and temperament, went just as far as the others would let him go.

All this helps to explain why Mattei, who was considered to stand on the left of the Demochristian Party, was less than happy when the 'Opening towards the left' was about to be realised; it was expected to lead towards greater strength and consistency in government as distinct from the anarchic conditions which short-lived cabinets without a policy and with hardly any popular support had allowed to develop. Under effective government the opportunities for Mattei's free-lance operations would have been seriously curtailed.

Actually, the experience of, say, France and Britain confirms my earlier contention that political power of the bosses of public enterprises can be and is successfully curbed by a strong civil service with its deep-seated traditions of putting the fundamental interests of the nation above sectional and ephemeral political trends. In pre-de Gaulle France the weakness of parliament and its lack of a coherent and consistent policy made the position of the *Administration* even more central and vital, since it provided a barrier to would-be builders of personal empires of the Mattei type. In Britain it was not only the continuous influence of the Civil Service but also the living power of Parliament which kept the headmen of public enterprises securely in their respective places, on the other hand there was no scope for prophets or visionaries either.

How do leaders of public enterprise get there in the first instance? The most primitive answer is obviously: political patronage. In fact there is a story dating from the Fascist era that the initials of AGIP (which, as will be remembered, stand for *Azienda Generale Italiana Petroli*) were really indicative of what it should have been called: *Azienda Generale Infortunati Politici*, that is – General Enterprise for Shopworn Politicians. From Mattei's AGIP and then from ENI such a joke could not have emerged; for all his instinct for keeping in with the useful trends, and his tendency to balance his team notwithstanding, he took the work which was to be done too seriously to give jobs to political protégés. Also any overriding loyalty to some party or other might have distracted his men from their primary allegiance to himself, and that, he must have felt, would never do.

Beyond Mattei, the danger of political patronage being prevalent in public enterprise is so obvious that it does not seem to have materialised on a large scale anywhere – except in Austria where it was part of the system itself. Even when and where the men at the very top of a public enterprise are straight political nominees and go on acting as such, their elbow room and sphere of influence are more often than not restricted by the professionals who really run the business. Actually the upper echelons of public enterprises all over the world are recruited by the same system of self-perpetuation of management which obtains in large-scale private enterprises. Indeed, with the biggest companies progressively taking on the character of institutions, recognising responsibilities beyond those to their shareholders, and with the public enterprises becoming more self-centred, the gap between private and public enterprise is narrowing, and their controversies will soon take on the character of the proverbial pot calling the kettle black.

Looking back once more at the man in the centre of the stage and at the background problem of public enterprise, the shape of things can be visualised.

Firstly, Mattei, a strong, almost demonic personality, who would have left his mark in any sphere into which he might have been drawn, does not fit into any neat pattern. Yet the fact that at a certain juncture he became almost single-handed the instrument of a trend towards independence from big business and from foreign control which could be got going only with State backing proves that this trend needed and could make use of the momentum of his character.

Secondly, the amphibian character of public enterprise, which has to orientate itself according to two different, not always compatible, goals, makes it impossible adequately to define its character; in fact its position remains anomalous in all circumstances against the background of private enterprise. Yet, since our economy is inevitably a 'mixed' one, public enterprise as a complement of and an antidote to the private one is now an essential part of our pattern. It can provide, as the case may be, a desirable stiffening or, at the other end of the range, it may engender a degree of mobility which would not be obtainable otherwise. It takes all sorts to make a world.

11   Topical Problems

This 'topical' was written in the aftermath of the June 1967 Arab–Israeli war. Obviously there was acute potential concern, threats were widespread (the Arab Summit which returned the Arab world to a new normality did not take place until August in Khartoum) and the atmosphere was tense, but as it turned out none of the fears expressed by Frankel in this piece in fact came to pass in the time-scale implied. Companies were not on the whole affected (apart from the nationalization of marketing interests in Algeria); OPEC two years later was stronger than ever it had been; energy diversification did not really get going for another five years; and oil consumer governments produced little in the way of new policy. In a longer historical context, the 1967 war could be seen as the beginning of a new phase of oil producer power but in this piece Frankel jumped the gun.

Elsewhere in this issue we deal extensively with the actual current repercussions on oil of the crisis in the Middle East. Here, however, we should like to list some of the likely longer-term effects.

(1) Since the Suez crisis of 1956, the political insecurity of Middle East supplies of oil to consuming countries has been at the back of most people's minds, yet with every year that passed the tendency to avoid drawing upon and developing higher-cost sources of energy has increased, which has meant rising reliance on Arab-controlled supplies: there is no doubt that even with the extremely high taxation in oil-exporting countries, most Middle East and some North African oils cost less at the point of consumption than practically any other available source of energy. The rude shock which has once again been suffered internationally will undoubtedly have its repercussions; in the first instance it will probably delay the

May/June 1967.

rundown of the European coal-mining industry, will encourage acceleration of atomic energy programmes, and will give further impetus to the search for and production of European natural gas. It will also cause governments and oil companies to favour diversification of their supplies in favour of non-Arab countries of origin – as far as Europe is concerned this will benefit in the first instance Venezuela, and in the case of Japan it will encourage the flow of investment capital to areas other than the Middle East. Although the problem of Iran is different because its cost factor has been no barrier, it can be assumed that that country will expect preferential offtake considerations in future, in view of its attitude in the current crisis. From the Arab point of view these developments are tragic, because the purely emotional and even in the short run ineffective endeavours to use oil supplies as a political weapon will have damaged their immediate and long-term chances very considerably.

(2) This problem has not yet been mentioned in any publication, but it stands to reason that the OPEC organisation, as we now know it, will have been seriously weakened by the current crisis, even if it should actually survive it. There have been behind the scenes all along some considerable divergencies of opinion between Venezuela and Middle East countries, simply because their problems and attitudes were determined by their respective different geographical, technical and economic positions. Within the Middle East countries there was, however, also a continuous tension between Iran and the Arab countries, partly due to divergencies in their political attitudes. There was, even before the crisis, the potential problem of Egypt's application for OPEC membership, which was to be made once Egypt became a substantial oil exporter which was expected to be the case in the near future. This might have upset the current balance of power within OPEC and made the position of Iran more delicate.

In the present circumstances one finds it difficult to expect the countries which have helped to make the Arab political boycott less effective to sit cooperatively alongside the Arab representatives. Also the whole status of OPEC, which was to our mind a reasonably effective 'trade union' type pressure group and which continued to have limited but, in the aggregate, worthwhile successes by haggling with the oil companies, does not stand up to a situation in which some (though not all) of its members may be striving for a fundamental shake-up; i.e. if the possible developments, to which we refer lower down, materialise at least partially, the whole balance of power will have changed and the type of operations for which OPEC has become known may be no longer relevant.

(3) Whereas it is our opinion that the immediate repercussions of the crisis on oil supplies will not be particularly far-reaching – even if the Suez Canal should remain blocked for some time, simply because it is already clear that the Canal is much less important now than it was in 1956–1957 – events may affect in a more permanent way the status of the Anglo-American oil companies and all those of other countries. It can be expected that the suicidal immediate measures of some Arab countries will be rescinded as soon as this is compatible with the inflamed local public opinion. On the other hand, the operating companies, affiliates of U.S. and British companies, are a sitting target and could be easily used as scapegoats in a way which would simultaneously satisfy more radical domestic political factions. We do not consider it impossible that there will be some endeavour to eliminate them as local operators and to call in other countries, without altogether disturbing the position of the original concession-holding companies as offtakers of oil. To reckon with such possibilities does not mean assuming that either the Italians or the French would be keen to step into the shoes of U.S. or British companies, nor do we expect that the Russians will altogether take over the industry. But there might be eventually elements of all these possibilities. We should not forget that the solution of the Iranian crisis of 1951–1954 did not involve simply the return of Anglo-Iranian (B.P.), which was politically unacceptable, but the formation of the broader-based Consortium more acceptable to the host country.

There must obviously be considerable revulsion in oil industry and Western government circles against yet another impingement on acquired rights, yet it is a matter of opinion whether the rôle of the producing affiliates of British and

American companies is really still an asset or has become more of a liability to companies and consumer countries alike.

The fact that the Western interests had a local position in the producer countries to defend made them less flexible in their decisions and more dependent on certain political pressures than they would have been if their rôle had been that of offtakers, for which they are irreplaceable. One could even imagine that it would pay the consumer countries themselves, who could make their bargaining position as vital buyers of oil felt more effectively if they had not hostages on the other side. The point may have been reached where the existence of Western-controlled companies in some of the producing countries may have become more of a liability than an asset. It may in the longer run be even more acceptable for the oil companies themselves – especially those like Shell, who are not very strong as oil producers in the Middle East – to be able to drive a hard bargain on their own account.

One of the reasons why we are somewhat doubtful of an immediate and complete restitution of the pre-crisis situation is that it might be difficult for American and British personnel to function effectively in circumstances which made possible the partly very ugly and frightening local situations which developed in the last few weeks. Although the company loyalty of the expatriate personnel cannot be doubted, there is probably a certain limit to what people can be expected to put up with.

(4) It is obvious that the current supply difficulties and any possible long-term structural changes would affect some companies more than others, positively or negatively as the case may be. It stands to reason that companies whose profitability was based principally on their availability of low-cost crude, e.g. B.P. and Gulf, would be adversely affected to a higher degree than companies who were always more dependent on the success of their downstream operations, e.g. Shell, or who were so widely diversified that no single event could affect them excessively, e.g. Jersey Standard.

Another consideration is that companies who are strongly involved in Venezuelan affairs, i.e. mainly Jersey and Shell, might derive some benefit from a higher level of output, which could possibly be achieved in that country. To a certain extent B.P. may also benefit with its 40% in Iran.

(5) It will probably be inevitable that some consumer countries will, perhaps for the first time, realise fully that a very large part of their oil supply system is entirely geared to American and/or British companies, and that they may be caught up in political problems of the U.S.A. and the U.K. in which they would otherwise not be involved. Although at a time of crisis the international versatility of the very big oil companies tends to be highly appreciated, this time the possible drawbacks are also beginning to be seen. This might in the long run strengthen the tendency of consumer countries to keep at least a sizeable part of oil industry activities in their own hands or in the hands of their nationals so as to give them direct access to world sources of supply.

(6) Provided that the status of international companies as actual producers of crude oil were to be eroded, with the elimination of some super profits and tax advantages derived therefrom, it might put an end to crude oil profits 'subsidizing' downstream operations (refining and marketing) in consumer countries and might bring about, as time goes on, a level of downstream margins capable of making these operations and investments an economic proposition per se, which they have not been for some considerable time.

12　Essentials Revisited 1968

Essentials of Petroleum *was republished in 1969 with a welcoming foreword by Professor Adelman (also reprinted here) and an epilogue, 'Essentials Revisited 1968', by Paul Frankel. In this epilogue (here in a shortened form of the original) he found that the* Essentials *were essentially unaltered. The jargon had changed somewhat, so that 'gasoline' replaced 'motor spirit' throughout and the 'incremental barrel' made its appearance. But the bones of the book were still firmly in place with their joints unimpaired; as Adelman said: 'This book deserves reading not as a pious exercise, but to gain understanding.' The epilogue also provides a useful reference to the main changes that Frankel saw in the industry in the intervening twenty years.*

Foreword

Having read this book many years ago and profited greatly thereby, I rejoice to see it re-issued. The style and wit would be striking even if English were Dr. Frankel's native tongue. Equally obvious is his first-hand experience. He might have written a once interesting now forgotten commentary. But his purpose was to reduce the vast detail of oil operations to a few simple theses. That is the method of science. The organizing principles are: competition and monopoly; the interaction of costs, prices and production; economies of scale. These, and not the picturesque detail or even the flash of insight, are what endures.

Since then, the world industry has expanded manifold, and its center of gravity has shifted to the Eastern Hemisphere. Moreover, our knowledge of oil production has itself been transformed as cookbook recipes have been absorbed into a body of systematic knowledge, reservoir engineering. Our

Epilogue to New Edition of *Essentials of Petroleum*, Frank Cass and Co. Ltd, London, 1969.

concept of its economic nature cannot but be affected. Yet the book is not obsolete. Those who have disagreed with the writer are – or should be – the most grateful for his having first blazed the trail through anecdotal irrelevancies to principles. This book deserves reading not as a pious exercise, but to gain understanding.

M. A. Adelman
Cambridge (Mass)   Professor of Economics
July 1968   Massachusetts Institute of Technology (M.I.T.)

The basic argument on which this book was based has stood the test of time: the oil industry's main features – its low price elasticity of supply and its limited price elasticity of demand – have continued to result in an industrial structure in line with its inherent needs.

On the supply side the extreme relation of fixed to variable cost persists. As a matter of fact, in the last twenty years a number of other industries have progressively adopted High Investment/Low Variable Cost patterns and my oil-industry type of analysis has become relevant for a substantial part of contemporary industry.

The high fixed-cost element of the refining, transport and marketing stages is reinforced by the risk factor in exploration and production. In this respect the sum total of one's efforts, much of which have led nowhere, takes on the character of fixed cost of the successful ones; this is accentuated by the fact that crude oil and natural gas exploration/production is exactly the reverse of coal mining ventures. Whereas it is comparatively easy and thus cheap to locate coal seams, it is very costly to bring coal to the surface; it is difficult and thus expensive to locate petroleum hydrocarbons in the subsoil, but once detected by the drill they tend to rise to the surface under their own steam, as it were, and in quite a few fields even without a great number of wells having to be drilled.

Consequently there is an overwhelming inducement to re-coup one's investment to the maximum possible extent and as quickly as possible since, within certain limits, the prime cost of every barrel of oil is low and, in many cases, lower than that of the previous one – a case of decreasing costs. These circumstances and not just the, by now forgotten, Law of Capture influence the oil producer at all times.

The motivation of the initial phase of the oil industry which influences the behaviour in the subsequent ones – refining, transport and marketing – now known as 'downstream operations' – is enhanced by similar ones affecting them directly: the analysis of refining cost is still valid today and the 'last 10 percent of a refiner's capacity' have since been christened the 'incremental barrel'. Only one thing has become clearer since: because it takes only two years to build a refinery but, say, ten years to develop a market position, especially for gasoline, and because it may take an infinite amount of time (and money) to find the oil, refining – in contradiction to the historical position – is not now 'a bottleneck': it provides no point of control; countries where such a status seems to be attached to refineries are those with a refinery licensing system and there it is the licence which matters, not the plant.

Backed by the experience of the inter-war situation, which was dotted with cartels which culminated in the As Is agreement, I was thinking in terms of combinations of operators to achieve market stability. The postwar extension of U.S. Anti-trust conceptions to the world at large, which was due to the comparatively new doctrine that U.S. companies had to comply with them also in dealings which did not touch the U.S.A., and also to the adoption of similar tenets in several other countries, had its distinct effects: my verdict that 'cartels and trade associations are the middle-class version of trusts' proved correct. The postwar period did not generate new spelled-out 'Combines', to use a now defunct term, but a set of conditions obtained which for some considerable time were extremely favourable for a limited number of very large companies. I shall presently deal with these developments, here I only want to lead up towards a short résumé of the compelling economic motives for concentration in a high-risk and high-fixed-cost industry.

If you have one oil well into which you have put all the money you have got and it is a 'gusher', you have hit the jackpot and are bound to get very rich. If, however, as is more likely than not, such a solitary well is dry, you are bankrupt and don't get a second chance. If, however, you have ten wells

– or better still a hundred wells – preferably not only in one field or in one region, or in one country, the odds are that, even if you know your business, a substantial number of holes will be dry; but few will be economic producers and, if you are lucky, one or two will be outstanding. Obviously the failures and the middling wells dilute the profits gained from the best ones, but it is obvious that this *diversification of risks* greatly increases the chance of survival, i.e. the future opportunities for profitable operations.

This spreading of risks is in line with the age-old principle of assurance companies: no one is sorry to have paid the insurance premium even if the incident against which protection was sought has not materialised.

This spreading of risks, vital in the aleatory phase of exploration, is also a substantial advantage 'downstream'. If you have only one refinery, or market only in one locality, you may be irremediably and fatally affected if economic and/or political conditions there turn out to be unfavourable for any length of time. If, however, you operate in a number of places, or countries, or regions or hemispheres, the odds are that adverse conditions at one point are balanced by better results at another. By being able to *average out* the bad with the better you manage to remain in business long enough to be still there when the tide turns in the trouble spot.

Thus the emergence (by growth and agglomeration) of very large economic units in the oil industry is due to their capacity for survival in the face of local difficulties – this is safety in numbers.

This problem of safety by way of diversification has yet another dimension: by operating on more than one level of the industry, i.e. by integration, the investment in each one of them is made more secure.

There are in fact two aspects of the security for which the operators, seeking full employment for their investment, must needs be striving:

There is first the element of assured outlet for the supplier and of assured supply for the offtaker, but there is also the advantage of being able to average out the profitability of, respectively, production, refining and marketing (and of transportation where applicable). It has been the experience in the U.S.A. (and for different reasons in the Middle East) that the producing phase of the industry provides a higher return (after tax) per dollar invested than do the subsequent phases of the industry; in France, for reasons inherent in the government-determined structure of its oil industry, refining is a more rewarding affair than is, for instance, marketing.

It can be said that an enterprise involved in all phases of the industry has a greater prospect of economic survival than have those working on one or two levels only. Hence, in our day, the rising tendency for international producers to absorb national refiners and marketers. This is an almost inevitable process when a crude oil producer can offer the refiner-marketer's shareholders a good deal more for their equity than would be justified by the profitability of the enterprise to be taken over. This becomes possible if the acquiring company puts into the calculation some of the crude oil profits which might never be made in the absence of the incremental outlet which control over the 'downstream' position could provide. Hence the saying that such companies could not survive on their own since they were worth more dead than alive.

I shall be dealing later with the more recent repercussions of the structure of the oil industry on the price problem and of the influence brought to bear on it from the outside. This up-dating of the original findings will best be carried out when the current policies of and for the industry are being reviewed, but we must first look at the influence the nature of demand has had on the market structure for oil products.

I stated that demand for oil had but limited price elasticity, and the description of the repercussions of price *changes* on gasoline demand still holds good; as the, now still higher, excise taxes on motor fuels prove, large-scale consumption is not precluded by a high price of gasoline – witness the rapid recovery of demand even after massive tax increases. What is true, however, is that the longterm level of prices has some considerable influence on the size of engines and therefore on the aggregate of gasoline demand.

For non-gasoline products – which were, a quarter of a century ago, of but secondary importance – the situation is entirely different. Whereas gasoline had 'no serious competitors' but was on the other hand confined to its preordained market and could grow only with it, the heavier parts of the barrel of crude – middle distillates and residues – had at their doorstep the vast market for home heating and industrial fuel hitherto covered by coal. Although the amenities of using a liquid played a significant role – hence the early progress of oil bunkers in the face of lower coal prices – the great leap forward in the postwar period was mainly due to the fact that (outside the U.S.A.) coal prices, being wage-determined, went up and up, whereas oil, helped by the stupendous discoveries of low-cost crude, became progressively cheaper. Thus we can say that for part of the barrel a high degree of price elasticity can be envisaged – at least it exists during the conversion period from one source of energy to the other. Once that conversion has been carried through, the competition of alternative fuels, e.g. in home heating, becomes less immediately relevant. The position is different though where dual-firing systems have been installed and there (mainly at the East Coast-U.S.A. utilities) relative prices play a great and almost instant role in the market pattern.

Since 1945 the process of substitution of coal by oil has gathered momentum almost everywhere. Outside the U.S.A. it was accelerated by coal becoming dearer as time went on, since the higher wage bill could only to a limited extent be compensated by technological progress, whereas oil got cheaper due to the discoveries of crude oil reserves whose costs were but a fraction of what they were in the U.S.A. on whose prices world market quotations used to be based.

Now, however, oil is no longer the Last Frontier of progress in the energy field; natural gas and atomic energy are growing faster and oil – especially fuel oil – is now in a defensive posture to which we were not used hitherto. On the other hand, the chemical industry has become altogether petroleum borne and is now an integral part of oil industry planning.

The transport business remains, however, the 'heartland': the railways have joined road transport as oil's virtually unsubstitutable market and there is as yet no clear sign that an alternative power for propelling vehicles and aircraft will be developed for some considerable time to come.

The fact that the refinery turns out what is called 'joint products' is still almost as relevant as ever but the difference in the realisations between the three main products has shrunk: gasoline is no longer quite the premium product it used to be and the 'discrimination' which loomed large is no longer what it used to be. The smaller overall margins are, however, acceptable because of the economies of scale which the industry has been able to achieve.

The item on 'where to build refineries' proved to be prophetic: after 1945 there was a rush to build refineries in consuming countries; governmental influence was prevalent in some cases but the industry itself soon found out that, once all parts of a refinery's yield could be placed locally, it was economic to build it as close as possible to the ultimate customers.

Whereas the basic approach of the book stood the test of time all along, the more political aspects of the picture looked badly out of date some years after it was written; yet most of it is now once again remarkably topical: *chassez le naturel, il revient au galop*.

My expectation that 'patterns for oil peace' would be drawn up by governments because oil was too serious an affair to be left to oilmen was not realised at the time. The International Petroleum Commission to be set up in accordance with the 1945 version of the Anglo-American 'Agreement on Petroleum' never saw the light of day because the Agreement was smothered by Congress, the 'independent' sector of the U.S.A. oil industry being afraid of 'the spectre of Federal control'. It was concerned about the danger of an international agreement having eventual repercussions upon the domestic scene, and little did its spokesmen know that some ten years later it would be they who would clamour for Federal protection in the shape of oil import controls.

Since nature abhors a vacuum, and since government agencies withdrew from the oil scene when the immediate postwar enthusiasm for international organisations began to wane, the big oil companies which then were the only organisations which covered effectively the flow of oil across the borders of countries were left to do what needed to be done, a function which was in keeping with the 'dual role' which they had been seen to fill. It may well be that, had the United Nations developed into an effective kind of authority, it could have tackled the task of setting up a viable international oil structure, but this was not to be.

Although even the biggest oil companies had in 1945 felt that their efforts needed to be underpinned by governmental programmes, they saw, soon enough, that they had been unnecessarily timid and that they could manage very well on their own. The desire of the American and the British Governments to avoid an 'oil war' was fulfilled in 1947 by a number of transactions in which the companies who had more low-cost crude oil than they could market themselves – Anglo-Iranian Oil Company Ltd. (now British Petroleum Company Ltd.) and Gulf Oil Corporation, Texas Company (now Texaco Inc.) and Standard Oil Co. of California – made some of it over in one form or another to those who had the 'downstream' facilities but were short of crude: Standard Oil Company (New Jersey), Socony-Vacuum Oil Co. Inc. (now Mobil Oil Corporation) and Royal-Dutch/Shell. Some of these transactions were between U.S. corporations but most of the others spanned the Atlantic, as it were.

Had the companies with surplus crude had to fight their way into the markets and had those who were short of crude had to develop rapidly some sources of supply of their own, the world market would have taken on an entirely different character. As it was, the new equilibrium led to what I have called the Ten Golden Years in the course of which profit margins and return on capital were exceedingly high. The substantial price reductions of oil in international trade in the last ten years, at a time when the prices of most other commodities have risen, were due to some extent to the inherent competition among the large international companies which were bent on increasing or, as the case may be, on maintaining their respective market shares.

Yet, the moves which eventually started to upset this privately organised equilibrium originated in governmental quarters: France had had a *dirigiste* tradition since the days between the wars: once again and with increased vigour it fostered state-backed French enterprises, whereas Italy – especially in the days of the remarkable Signor Mattei – set up a string of state-owned companies as a means of matching the strength of (foreign) international oil companies. India, Paki-

stan, Ceylon and later Japan, took steps to influence investment and pricing policies of the oil companies; in all cases it
emerged, as had been witnessed earlier on in the U.S.A., that
the concomitant of Big Business was Big Government: since
the international oil companies were such large units there
could be no countervailing power on the business level in any
one country; only its government was a unit of equivalent size
and weight.

The scene shifts to the U.S.A.: the enormous difference in
cost between most of the indigenous crudes and those from the
Middle East and (to a lesser degree) from Venezuela, would,
if price alone had determined the flow of oil, have resulted in
foreign oil replacing much of the indigenous crude. The
onrush of Cheap Foreign Oil was, however, held by that kind
of 'invisible hedge' which had started to grow even before
World War II. The fact that most of the low-cost foreign oil
was controlled by a few companies who had a vital stake in the
U.S. domestic market made it possible for them to apply in
the first instance self-interested self restraint or one could at
least for a while rely on what an American once called the
'collective common sense of the several competitors'. Such
so-called Industrial Statesmanship began to wear thin in the
late '50s and the U.S. authorities were eventually driven to
imposing mandatory import controls without which the internal system of market-demand proration, applied in the
main oil-producing States, could not have been maintained
any longer.

Thus the U.S. had become a market which was subject to a
high degree of governmental control, whereas American entrepreneurs abroad kept on demanding freedom for action
then and there. Such freedom became anyhow qualified by
forces coming from other quarters making themselves felt:
when increased competition eventually drove Middle East
prices down from their erstwhile high level – high in comparison with its cost of production – the producer-country governments, understandably concerned about the threat of their tax
'take', joined forces and, in 1960, formed an Organisation of
Petroleum Exporting Countries (OPEC) which made it possible to institute 'collective bargaining' which in some respects
proved much more effective than could have been any

technique of individual negotiations.

Most oil producing countries – following the lead of Iran – have by now also established national oil companies of their own, which, apart from handling all or most of the domestic refining and marketing business, have in some cases entered the spheres of exploration, of production, of refining and of export. It is yet too early to gauge accurately what (if any) fundamental impact the emergence of these national companies will have on the fabric of international oil trade, which is still mainly managed by a limited number of substantial companies with a worldwide scope.

The consumers, or rather the oil importing countries, do not always fully appreciate that the oil companies together with the producer countries may present them with the prospect of high-cost energy; indeed, for some time in the past those importing countries, which had a high-cost indigenous source of energy to protect – coal in Europe and oil in the U.S.A. – appear to have fancied a situation which did not shake their own vested interest too much and too soon.

Looking at the situation as a whole one sees a form of symbiosis of the two systems, of corporate enterprise and of governmental planning; or should one perhaps, with equal justification, talk of corporate planning and governmental enterprise? No longer does it make sense to consider the 'private' or 'free' enterprise to be subject to 'interference' by governmental authorities; one could say with as much (or as little) justification that corporate activities 'interfere' with national policies. The truth is that neither system is self-contained and that they depend on each other for their respective developments. Whereas this is obviously a generally applicable concept it has some specific relevance in and for the oil industry.

The international companies, who look for oil where they find it and refine and sell it where they may, play an indispensable role by virtue of their diversification. It gives them and through them the world stability by way of the insurance element mentioned earlier on and flexibility to adapt themselves effectively not only to peaceful change but also to the shock of war-borne crises. They do in fact attempt and to some extent achieve a worldwide optimization of effort and investment.

Yet they cannot have it all their own way, because of the incidence of local, national and regional circumstances and interests. If the whole world were one unit, if there were no local and especially national tenets which render unacceptable a degree of division of labour under which the low-cost sources would be preferred altogether to high-cost ones, wherever they are, optimization on a truly world scale would not only be desirable, it would be possible. But the world is not one unit, local industries on which the livelihood of large sectors of the population may depend cannot simply and rapidly be optimized out of existence and, still more relevant, national security in the political as well as in the economic sense of the word cannot always be subordinated to the possibly short-term convenience of low-cost imports. In fact there is something like a local, national or regional optimization which is determined by a number of elements germane to their particular situation.

What really matters is to find a viable and judicious blend of the two optimizations: if the world-orientated system is apt to destroy indispensable elements it has to be curbed, but if the locally or nationally motivated concept leads to a degree of cost increase which becomes self-defeating, it has to be moderated. Thus there *is* a target such as an overall optimization: at worst it might be a tenuous compromise, but at best it can be a mutual adaptation resulting in a pragmatic sort of harmony. Almost a quarter of a century after 'Essentials' was concluded, one should find as inevitable and as desirable as ever the idea of the coexistence of governmental and corporate policies.

13 The Current State of World Oil

*The late 1960s were a fertile period for analyses of oil producer/
consumer/company relationships. This piece, published in MEES (then,
as now, one of the leading weeklies covering oil developments, particularly those connected with OPEC and the Middle East) was one of Paul
Frankel's additions to the debate. It is easy enough to see why he gained
the reputation, in some OPEC circles, of being anti-OPEC, but a
careful reading of this article proves him to be no such thing. He always
followed arguments to what seemed to him to be their logical conclusions
and, if the end-result was unpopular with OPEC or the companies or the
consumer governments, so be it. In this article Frankel clearly shows
how, once the companies have established the principle that costs are
passed on to consumers, the level of tax ceases to be so important to them
(note that this was written eighteen months before the Tehran Agreement
as a result of which this principle became a lot clearer to many more
people). At the same time he is merciless in analysing OPEC's expectations for maintaining price by agreeing output controls and suggests that
state-company (and other) competition for concessions was probably
counter-productive in terms of national interests. He is clear in his
prescription for consuming countries to create their own organization if
they are opposed by a militant organization on the producer side. In other
words it is an article that is in no way self-serving but is designed to
create and advance the debate.*

The debate on the future of the supply of oil from a few
comparatively small countries, where most of its reserves
appear to be concentrated, to a substantial number of countries virtually bereft of indigenous oil seems at long last to
focus on the essential problems. If this should happen now it
will be to a great extent due to a recent statement made by
Mr. Ahmad Zaki Yamani, Saudi Arabian Minister of Petro-

leum and Mineral Resources, to those made by other Middle East spokesmen and in some way also to a presentation recently made in Boston, Massachusetts, to the Financial Analysts Federation by Mr. D. H. Barran, a Managing Director of the Royal Dutch–Shell Group and Chairman of the Shell Transport and Trading Company.

For some time past the interested public has had to be content with titbits such as those on 'expensing of royalties', so technical in nature as to be indigestible for people not directly concerned and for the rest the stage appeared to be occupied by a slanging match between oil companies and a producer-country government for the control of some concessionary acreage in Iraq involving nothing less than the fight for The Rights of the People and for Freeing of Natural Resources from the Yoke of the Foreign Monopolies. Then there were the peremptory requests of the Shah to increase – there and then – the offtake of oil from Iran, regardless of what happened to other sources of supply and of course there is always with us the recollection of what took place in the wake of the 1967 Six-Day War.

In the midst of all such false trails and lingering memories there appears Mr. Yamani who starts talking of the things that really matter and opens a discussion in which genuine analysis of a situation can find its place. Whether one agrees with Mr. Yamani's stand or feels compelled in some respects to put up countervailing arguments, one has to recognize gratefully that Mr. Yamani and a few others are about to raise the discussion above the traditional level: the rest of the world can rejoice in having in the Middle East, at long last, what the French call *interlocuteurs valables*.

The presentation of Mr. Yamani falls under three headings: the Six-Day War and its aftermath; the objects and the means of price maintenance for oil; the relationship of oil producer countries to oil importing countries and to the oil companies.

I

In respect of the first-named item, Mr. Yamani's statements are remarkably candid, he admits that the decision to boycott the U.S. and Britain '. . . was a wrong decision taken on the

basis of false information.' He goes on to say: 'It hurt the
Arabs themselves more than anyone else, and the only ones to
gain any benefit from it were the non-Arab producers.'

So far so good – Mr. Yamani's statement reinforces the
general feeling that withholding of oil supplies in order to
bring political pressure to bear on consuming countries having been a signal failure is not likely to be tried again as long
as everybody concerned has the 1967 events in mind. This, I
suppose, is on a par with the idea of all-out nationalization of
oil concessions never having recovered from Dr. Mossadek's
debacle in the early fifties.

Yet consumer countries, having been threatened twice
within a decade, with Mr. Yamani's country featuring in both
cases, will not be altogether reassured: on the level of economic long-term considerations one could assume that rational
thinking would preclude future adventures – but can such
attitude be taken for granted in all circumstances? Will the
regimes in the relevant countries be of the standard of enlight-
enment Mr. Yamani has now attained or will they once again
be swamped by an emotional upsurge? Mr. Yamani's welcome statement will not altogether manage to undo the trend
towards alternative sources of supply of energy even if they
should for the time being involve heavy investment and additional cost.

There is another relevant point: only a year ago the futility
of the existence of affiliates of American and British oil companies in Saudi Arabia or Kuwait was blatantly exposed:
whilst they were allowed to go on operating after the June
War, lest the inflow of funds be interrupted, they were precluded from supplying the very countries whence they had
come in the first instance in order to assure their oil supply.

II

Mr. Yamani's approach to the price and related problems is
the central point: he takes the world as he finds it and he takes
it for granted. It is apparently not for him to analyze how a
situation has come about in which – to use his own considerably understated figures – an exporting country is now able to
levy a tax on oil amounting to four times the cost of produc-

tion. There is in fact no precedent of a similar 'take' in respect of any other commodity, and it is thus perhaps fitting to scrutinize the means by which such 'take' is to be maintained and increased by reviewing briefly the circumstances in which it came into being.

The salient fact of the Middle East after World War II was that its low-cost oil was eased into a world market geared to much higher-cost crude. At first the quantities of the 'new' oil were too small to determine the going price, and when the volume rose, the interest of the countries with higher-cost energy sources – U.S.A. (later also Venezuelan) crude oil and European coal – not to see them be pushed out by the new low-cost oil too quickly, and too far, coincided with that of the international oil companies; most of these also had high-cost supplies of their own to protect and were anyway bent on making hay whilst the sun shone; thus all concerned were apparently happy to inhibit the inherent tendency for low-cost oil to expand rapidly by underselling its higher-cost competitors. After a few years, i.e. when the profit-sharing (50/50) concept became prevalent, they were joined in this pursuit by the producer-country governments.

This situation obtained for what I have called The Companies' Ten Golden Years, but in the decade after 1957 there developed a progressive erosion of the oil industry's margins, due to the slow but inexorable working of competition in a 'long' market in which, in most instances, the additional slice of production was the one with the lowest cost. It became then, within reason, perfectly rational for operators who had much greater reserves than they could possibly utilize within the time span of their concessions to extend their sales even at the expense of their margins. The established companies progressed on this path but warily and I have previously talked of a kind of 'invisible proration' practised by them. That on the other hand some of the newcomers developed their resources more rapidly was equally understandable; everybody – including governments such as that of Libya – appreciated that their barrels had to have at first a greater velocity, as it were; as a matter of fact, such procedure was in line with the Discovery Allowables of some U.S. proration schemes which are higher than those of established reservoirs.

Until fairly recently the producer-country governments had no direct influence on the volume of the oil exports from their country. Such decisions were in the domain of the operating companies and in fairness one has to acknowledge that it was in the first instance the oil companies which created a setup in which the low-cost oil was sold at prices yielding an outsize margin of profit: the producer-country governments were presented with a fait accompli although some of them in the Middle East might actually have benefited from a higher output at lower prices; since the companies – with London and Washington as sympathetic onlookers – opted for (comparatively) limited output, producer countries, bereft of the opportunities of more rapid growth, were perfectly justified in claiming a substantial share in the resulting high margin per barrel.

It must be borne in mind, however, that what started in the 50/50 setup as a genuine share in a margin has since 1960 (i.e. since OPEC, and since the virtual elimination of 'realizations' as a tax base) become a straight cost item: *an income tax has been transformed into a levy* and Mr. Yamani is not alone in stressing the importance of the posted (or tax reference) price which provides a floor for the price of oil.

I have talked of the fortuitous constellation which some twenty years ago brought about profits on low-cost oils which were in fact an 'economic rent'; as far as the oil companies were concerned this has by now been whittled down, first by the governments' profit shares and then as we have just seen by the working of the forces of competition. The 'take' per barrel of the producer-country governments, however, remained virtually intact: the cut resulting from the 1960 reduction of posted prices in the Eastern Hemisphere – the 1959 reduction made good mainly the earlier Suez-crisis increase – has been practically wiped out by the subsequent arrangements including those for expensing of royalties, and in Venezuela the tax element per barrel is now about as high as it ever was except during the aftermath of the first Suez crisis.

This inviolate state of the governments' tax take has been due partly to their peculiar relationship to the oil companies operating in their territories, about which I shall have something to say towards the end of this article, and partly to the fact that hitherto the producer countries were not faced with world market conditions but had only a bilateral relationship with one or more concession-holding oil companies whose prerogative it was to determine how much they would actually produce and export. Consequently the governments were spared the classical dilemma of the industrialist – whether to stress the volume of production even if this meant a lesser realization per unit or to concentrate on margins, possibly at the expense of volume. Since it was not for them to decide upon quantities to be produced they could happily concentrate on trying to scale up their benefit from whatever the companies could or would produce and export. Mr. Khalifa Moosa, the Libyan Oil Minister, said only recently: 'The companies are alone responsible for market (conditions) and their intense competition among themselves is to be blamed for any deterioration in their position. We as producing and exporting countries cannot be held responsible for what we do not actually take part in.'

All this is likely to change gradually if and when the producer countries increase their control over what goes on: by having oil other than royalty crude, which they will tend to dispose of through their national companies, and by influencing the concession-holding or contracting foreign companies in respect of their offtake programs, these governments are becoming a source of industrial decisions. Iran actually now provides examples on both these levels.

It has been for some considerable time Professor M. A. Adelman's and my contention – he getting there by way of economic analysis and I by way of oil industry experience – that once the oil exporting countries have achieved self-determination, as it were, they will – in the absence of effective arrangements to the contrary – behave the way the oil companies did when it was their turn: if you make seventy cents to a dollar a barrel on oil that it takes a fraction of that amount to produce and whose production can be raised without substantial effort and at (in many instances) decreasing cost, you will try to maximize your profits by increasing the sales volume although it may mean lower unit realizations, which, at first, would mainly affect incremental quantities.

Mr. Yamani is obviously well aware of this problem and he

actually said in his Beirut lecture: 'Some exporting countries –
although they may not always be members of OPEC – appear
to think that if they see an opportunity to make a little extra
profit on the margin with incremental sales, it is fine to go
ahead and do it. But the fact is that this little "extra" profit
will inevitably result in a reduction in the bulk of the profit
accruing to them, us and everyone else.' Did he also think of
the Kuwait law which gives its national oil company KNPC a
rebate equivalent to 80 percent of the government's tax and
royalty income on each barrel of Kuwait-origin crude exported by KNPC (as refined products) to a 'new market'?

This leads us to the crucial problem of effective control of
the quantities to be exported by each of the countries involved: as long as they could confine their endeavors to ex-
tracting better *terms* from the concession-holding companies
they had identical or at least parallel interests, hence OPEC's
success in that respect, but when it comes to *quantities* the
governments are shown up as what they in fact are, straight
competitors of each other; thus the coordination of the several
interests is subject to the same problems which have beset
would-be cartel makers since time immemorial. If Mr. Yamani were to consult the literature on the subject he would find
ample evidence for the fact that price cartels which are not
solidly underpinned by quantitative arrangements, i.e. by
well-policed quotas of one kind or another, tend to fail in due
course; in fact the higher the benefit derived from the cartel
price, the greater is the temptation to shade that price; such
bushfire once it has started more often than not leads to a
general conflagration and eventually to the breakup of the
cartel.

It is not surprising that the idea of international proration,
assiduously canvassed within OPEC earlier on, never really
got off the ground: there was the difficulty of finding a generally acceptable method of establishing quotas, the conceivable
ones – proved reserves or historical positions or population of
the respective countries – being unacceptable to one or
another participant whose adherence to a program was yet
indispensable; also, the OPEC members are now too well off,
their 'take' too high, for all of them to see the writing on the
wall which would make them ready and keen to swop ambi-

tions for security: it is only after the fat margins have gone and when the potential members are fighting for their lives that the sacrifices inherent in a quota arrangement are made – cartels have rightly been called 'children of distress'; it may also have been shrewdly realized that a formalized quota setup might possibly call forth defensive instincts in the ranks of consumer countries and it was also clear that a classical form of proration would be altogether unacceptable to the international oil companies.

There are very good reasons for this: the strength the international oil companies have in the producing countries no longer, as it might have done earlier on, derives from their ability to find oil, to produce and to finance it; all these services are now readily available on a professional level. What does, however, matter is what I have called on previous occasions their Power of Disposal, i.e. their control of downstream operations, in the first instance in the refining and marketing spheres. The producer countries depend on the companies' ability to offtake regularly and in massive quantities and the companies have therefore a strong leverage as long as it is they who can decide within certain limits where the oil should come from. Due to joint ventures and swap possibilities, there is a certain flexibility in the programs and thus a means of favoring among the producing countries the just ones and keeping the unjust ones at bay, simply to make evident that 'crime doesn't pay'. In a formalized proration system based on determinants other than company policies, that leverage would disappear and with it much of the influence the companies can bring to bear. It would also make it more likely that consumer countries would query the companies' role as the providential purveyors of oil.

Since under virtually all main concessionary arrangements the decision as to the amount to be exported – over and above certain minima which were left far behind anyhow by the increase in demand – was the operator's business, a long-drawn-out campaign for a change in those agreements would have been necessary which would have detracted OPEC from its revenue-providing tasks; this, coming on top of all the other difficulties of international proration, made the whole idea virtually impracticable. If further proof was needed for my contention it would be provided by Mr. Ashraf Lutfi of Kuwait, lately Secretary General of OPEC, in Chapter VI of his book *OPEC Oil*, recently published in Beirut. His idea of Joint Programming would require a remarkable degree of compliance by OPEC members and of complacency by everyone else.

It is probably for all these reasons that in the July editorial of *Review of Arab Petroleum and Economics* OPEC is being implored that it 'should agree that they will never agree on a system of prorationing.'

Its Editor, Mr. Abdul Amir Q. Kubbah, seems to fear that in relying on the prospects of a proration system countries would 'sow the seeds of the future destruction of the price structure,' whereas he in fact wants them to become 'more price conscious.'

However, Mr. Kubbah has a remedy and with engaging frankness he suggests one should find 'the most effective way of fighting competition and promoting monopolistic tendencies.' For good measure he envisages that 'the oligopolistic structure of the oil industry should be strengthened by all possible means.'

Odd as all this may sound it is only a more blatant version of what Mr. Yamani thought in Beirut where in reply to a question he said that the major companies 'in ten years or so could turn out to be our natural allies.' Only a little later in the same session he seems to have realized that ten years was an unnecessarily long period after all, and he added: 'I agree that any alliance between the producers and the majors should start as soon as possible.'

The significance of this producer-country government approach is enhanced by these sentiments being quite obviously shared by oil companies. It is there where Mr. Barran's statement referred to earlier on comes in. He said on 6 May 1968: 'Pressure from the producing governments on costs is something that we can live with provided we are not at the same time denied freedom to move prices in the market so as to maintain a commercial margin of profit.'

This can only be read as meaning that some oil companies are in the last resort not averse to maintaining the crude oil tax level, which they had appeared to fight tooth and nail, or to allowing it to be increased still further, provided only that, in the absence of price control and/or other measures in consumer countries, they can get the commensurate realizations in the market place.

It is fairly obvious that henceforth producer-country tax increases, as long as they apply equally to all operators, do not particularly concern the oil companies since they have in the end to be paid by the consumer countries and are therefore only a transitory item in the companies' profit and loss accounts. (It was the fact that the established companies in Libya had for a while to pay higher taxes than the newcomers which was so objectionable to the former.) Mr. Barran himself had pointed out earlier in his speech that by the reduction of the so-called OPEC allowance 'other companies are similarly affected,' but it is difficult to see why he thinks that this is 'cold comfort.'

The idea that the incidence of some cohesion amongst the oil-producing countries, for instance through OPEC, was not altogether unwelcome to oil companies is not a new one. As early as February 1961 the *Petroleum Press Service*, published in London, which on such matters is not in the habit of straying far from the philosophy of some major companies, said in an item headed *OPEC Starts Work*: 'Short of international pro-rationing much can no doubt be achieved by the collective influence of OPEC, if only by creating a climate of thought, both among themselves and on the part of the competing oil companies, which will check any deterioration in the present highly sensitive competitive situation.'

There are a number of varied reasons why the oil companies should lean towards their link with producer countries: in that domain the prospects of success and deprivation respectively are concentrated and immediate, whereas any influence which the consumer side can bring to bear has so far been diffused and somewhat remote. One can get or lose a concession by the stroke of a pen, whereas the downstream reactions are rarely as clear-cut.

There has always been a good deal to be said in favor of letting the competitive forces work, and due to them oil has in fact steadily become cheaper as time went on. If, however, the operation of market forces should be seriously inhibited by a closer understanding among oil exporting countries, aided and abetted by oil company policy, the situation would be entirely different.

III

Before assessing the potential significance of what one of the shrewdest American observers of the oil scene has called an Unholy Alliance, it is necessary to analyze the current status of its component parts – OPEC and its members on the one side and the oil companies on the other.

OPEC was set up on classical trade union lines: like the wage earners of old, the exporting countries realized that they had to hang together or be hanged separately, that only the extension of any individual conflict to the whole range of alternatives, complete with 'sympathetic strikes' and with declaring certain supplies 'black', gave them a chance to match the built-in solidarity which the oil companies display in such matters: the traumatic experience of the Mossadek episode, when the other Middle East countries cheerfully let their production be stepped up, whereas the other oil companies helped Anglo-Iranian to adjust itself to the loss of its Persian crude and of its refining capacity, resulted in the key resolution of the first OPEC Conference in September 1960 (Resolution No. I/IV). There it was laid down that members were to stand by any member engaged in a conflict with an oil company recognized by a unanimous decision of a Conference. In this case, 'no other Member shall accept any offer of a beneficial treatment, whether in the form of an increase in exports or an improvement in prices, which may be made to it by any . . . Company or Companies with the intention of discouraging the application of the unanimous decision reached by the Conference.'

This OPEC activity, to my mind perfectly sound and legitimate, takes on a somewhat peculiar character only through its members being governments whereas its scope is altogether in the commercial sphere. Hence the tendency for multinational pressure groups to develop pretensions to the standing of an international organization – which yielded them, surprisingly enough, a kind of exterritorial status in Vienna. OPEC also makes use of the idea that it consists of developing countries whose interests it represents in confrontation with the rich industrialized countries, who are the main importers of crude oil. True enough, the fact that in this century big oil has been found mainly in hitherto underdeveloped countries, has tended to reduce the wealth-gap to some extent, but:

(1) Not all oil importing countries are rich nor are all oil exporting countries poor: today India and Pakistan subsidize Kuwait and Abu Dhabi and for that matter Brazil subsidizes Venezuela.

(2) The question can be asked whether a situation can be defended on grounds of international equity, as distinct from commercial opportunity, in which virtually the whole 'economic rent' to which I have referred, is being appropriated to a very narrow circle of mostly small countries: if the world owes them a living it would hardly be at the rate of getting on to 5,000 million dollars per year.

Hence my feeling that the emotional approach to the problems involved could prove a boomerang to those who, unlike Mr. Yamani, still use the slogans of yesteryear.

On a more mundane level, OPEC, being a kind of trade union secretariat, has not only to justify its keep, it must be seen to be doing so, and therefore cannot avoid keeping on putting up claims for the betterment of its members; thus the idea that it was a law of nature that (apart from the increased revenues derived from rising demand) the terms of trade should be shifting continuously and irreversibly in favor of oil exporters, is being kept up without too much regard to the real factors involved.

OPEC, of course, is not a monolithic entity and the latest Resolution (XVI. 90) emerging from its June Conference in Vienna – 'Declaratory Statement of Petroleum Policy in Member Countries' – shows how narrow the common ground is on which its members are prepared to stand. Yet, two features stand out in a rather flat landscape: Participation and Renegotiation.

I do not know whether the slogan of Participation of oil producer governments in the equity and/or in the operations of foreign-based oil companies owes something to the current political fashion in France, but in the oil sphere too the word may mean anything or possibly nothing: it is not even clear what Mr. Yamani had in mind when he asked for it, although the self-contained structure of Aramco – a company which distributes its profits by way of dividends and has no downstream business – would lend itself more easily to a genuine equity participation of the national government than would that of the Iraq Petroleum Company or the Kuwait Oil Company.

It may well be that nothing much is to come of such a campaign, but it might be considered as one more turn of the screw, a tightening of the vice in which the companies have found themselves for some time past. It could also be a step towards that 'alliance', participation providing the operational opportunities for an alignment of interests.

The other item in the OPEC resolution covers the right to ask for renegotiation of agreements if the oil company 'obtain(s) excessively high net earnings after taxes.' This looks like a new version of the earlier idea of Selective Profits Tax originally launched in Venezuela, but abandoned later. However the term 'excessive' should be defined the implication is clear: the government insists on a floor to its 'take' by way of a posted price which is not to be queried and beyond this the sky is the limit; the company on the other hand has no floor for its earnings but it has to accept a ceiling. Those who should let such terms be imposed upon them will provide the world with the unedifying spectacle of a man who rushes up a 'down' escalator.

IV

The general posture of oil companies is being determined by considerations which have so far failed to be clarified in public: whilst realizations for oil products and through them for crude oil have been going down over the last ten years, trimming margins apace, the terms of concessions have if anything become steeper than they were: oil is sold by the companies in a buyer's market but is apparently 'bought' by the same companies from the producing countries in a seller's market. Surely this is a phenomenon which takes some explaining.

One facet of all this is the difficulty of concession holders to refuse renegotiations of existing agreements when the Oil Minister can point to a long queue outside his office consisting of new applicants prepared to sign on almost any terms. The rationale of this seemingly irrational behavior lies in the extraordinary concentration of profitability in the case of the most successful oil-exploration ventures. The very uncertainty of the results which deters one type of operator encourages another, who is sometimes right in assuming that it will be he who will 'make a killing' when others have failed and foundered. This attitude is less reckless than it looks and the oil industry would not be what it is without a certain element of genuine gambling. The peculiar relationship of profit and loss in this sphere can best be explained by a proper re-interpretation of the concept voiced by Sir Maurice Bridgeman in BP's Annual Report for 1965 that 'one success must pay for several failures.' Statements to that effect have been made so often by oil company spokesmen that Sir Maurice seems to have stressed only the obvious, yet what he said is to some extent based on a fallacy. The statement is in fact correct for each individual firm but not for the whole industry – unless there prevails a state of monopoly.

Obviously, an operator whose successes do not balance his failures overall will sooner or later go out of business, but that is exactly what goes on happening in a competitive society. The inept or the unlucky ones disappear and this is a charge on the capital available to the community, but not on the oil industry itself, certainly not on the successful operators, who even benefit quite often from the highly-paid-for experience other people's failures have provided for them. Lotteries, incidentally, would not make sense if each participant would get his money back – the organizer has to get his out anyhow and the big win is possible only because quite a few lose their money: what makes lotteries attractive is the considerable difference between the amount at risk for each lottery ticket and what can be expected if one 'hits the jackpot'.

Whereas the successful producer cannot establish a claim to being recompensed for the losses of another company which has failed to find oil, the argument is correct for the sum total of investments and it actually applies to the investments the

industrial countries (a great number of them being oil importers) make in the producing areas. Thus, *what might be reasonable for any individual enterprise which takes into account the prospects
of a bonanza, may become a heavy and unjustified burden for the
economies and balance of payments of the investing countries for whom
the sum total of investments and tax liabilities is the price paid, i.e. the
'social cost' of the oil.* It is perhaps not too far fetched to envisage
that if the OPEC countries manage to establish minimum
terms for what for simplicity's sake we still call concessions,
the importing and investing countries might consider putting
a ceiling on the 'price' to be paid. They would in fact take an
interest in the terms their nationals would offer to producer
countries. They might be driven to so doing because the
aggregate of the economic assessments of the bonanza-hungry
entrepreneurs may set a standard of 'cost' which is not necessarily justified overall.

The proper analysis of what really matters had been inhibited by our terminology: in fact, the oil companies are not
only freelance operators, *they are actually purchasing agents of the
consumer countries* and it is not unreasonable to ask whether the
interests of these companies ever have been or at least are any
longer identical with those for whom they get the oil. The odd
thing is that, contrary to the proverbial situation, the one who
does pay the piper is the one who is now excluded from calling
the tune.

It is not surprising that Mr. Yamani dissociated himself
from my thinking and said that he would not wish to be
confronted by the companies and through them by the consumer countries as *buyers* and that he much prefers to deal
with the oil companies as *operators*. It is obvious that he
considers them, with their investments in the producer countries, as being easier game. The traditional homely expression
of Host Country will perhaps have to be transposed to Hostage Country; there is a saying that 'a million dollars is a
coward' – what would a billion dollars be?

V

It is now possible to draw the various strands together:

(1) The postwar advantages of low-cost oils, though still existing, are shrinking, and the incidence of the '$1 per barrel' tax burden begins to limit the expansion of oil, especially in competition with atomic energy.

(2) Price maintenance by way of posted prices, though endangered by volume-minded governments, may become a more likely bet if the mutual sympathy of producer-country governments and oil companies leads to a de facto setup in which the several interests are 'harmonized' in such a way as to maintain, possibly to enhance, tax take *and* profit margins.

Although such understanding would still be subject to problems bound up with any attempt to maintain prices and may not work perfectly, its comprehensive yet diversified and flexible nature would give it a more than reasonable chance of achievement in the course of a fairly long run.

(3) These circumstances gain in relevance because the producer interests have emerged as sovereign governments and as commercial entities rolled into one: any rights of an oil company resulting, as one would assume, in the first instance from a balanced give-and-take situation can apparently be changed unilaterally on grounds of sovereignty. It is interesting to see that those who coined the phrase 'a sovereign government does not negotiate, it legislates' have in fact attenuated greatly the relevance of *any* terms a government can offer to an entrepreneur; if the all-pervading bonanza were not the apparently irresistible bait – the fond hope that one Occidental *does* make a summer – there would long since have made itself felt a resistance movement of prospective candidates for concessions; they would have been 'voting with their feet.'

Although harsh views à la Tariki are not voiced by men of Mr. Yamani's sophistication and although OPEC talks euphemistically of renegotiation, the ultimate sanction of unilateral governmental action is a useful means of concentrating the mind of foreign enterprises.

The much vaunted concept of the countries' 'Permanent Sovereignty over their Natural Resources' – OPEC last June referred to U.N. Resolution 2158 of 25 November 1966 – seems to be based on a tendency to overestimate the sellers' strength and to disregard somewhat rashly the need to find a market for such resources in other countries. Failure to appreciate this need may make the producing country's sovereignty over its resources permanent indeed.

(4) At the time when the few established oil companies – the Seven Sisters – enjoyed a very substantial gross margin, say 70 cents per barrel on crude oil, it made perfect sense for other oil companies and for consumer-country governments to endeavor to establish direct access to 'cost oil'. By doing so they were to bypass a tollgate or turnpike manned by a bunch of profit makers and could hope to lay their hands on exceedingly valuable properties.

For several reasons things did not work out exactly like that:

(a) The profit margins of oil companies having been eroded by way of competition to roughly a third of what they were immediately after the introduction of the 50/50 setup, the *super*profits of the Old Gang had virtually disappeared and with them the main reason for trying to acquire one's own 'cost oil'.

(b) In order to obtain positions in producer countries the newcomers had to offer terms which, on the whole, made their oil (if and when it was found in commercial quantities) dearer than was the 'cost oil' of the established operators.

(c) The more onerous terms of certain new concessions tended to weaken the status of the old ones and increased in turn their costs and the level of terms for all new offerings (offshore Iran is a telling example) was also raised – consequently the importing countries had to pay more for all their oil imports than they would have done otherwise.

(d) Completely absorbed as they are in the endeavor to do now what might have been worthwhile twenty or even ten years ago, newcomers fail to realize that, in trying to

supplant the established oil companies they have to in-
gratiate themselves with the producer countries: in the
process of acting as a second-rate producer they forego the
opportunity of being a first-rate buyer and of applying
pressure where it could do some good.

It is my contention that much of this kind of activity
carried on, first in Italy and then in France and Japan, by
government-sponsored enterprises, far from achieving its
intended goals, might have been rather detrimental to the
national interest(s) it was designed to serve.

(5) I have no doubt whatsoever that countries which encountered the good luck to have a great deal of (exhaustible)
low-cost hydrocarbons in their subsoil are completely justified in making the best of them and the way their crude
oils have been introduced to the world markets has provided for them an enviable starting position which at least
the richer countries among oil importers should not be-
grudge them. There is, however, now no reasonable
balance of power, with OPEC, performing cartel functions
and buttressed by the executive power of governments, for
which the oil companies, inhibited by living under the
shadow of Anti-Trust regulations, are no proper match.

It is evident that in the circumstances only governmental policies in the consuming countries can provide a
modicum of countervailing power. I believe I have just
shown that one such line of approach – the acquisition of
'cost oil' – does not, or at least not any longer, fulfill the
task of reducing substantially the cost of imported energy.
Only by remembering what a smokescreen has almost
obscured, that in a market where potential supply greatly
exceeds current demand it is the buyer who is in a strong
position, can one devise an adequate consumer-country
policy.

Japan, however ill-considered may have been some of
its attempts to integrate backwards by scattering indis-
criminately and over a wide area its efforts to produce oil
abroad, has also had the right idea, previously developed
by the Indian Government, that it should take a hand in
securing acceptable terms for the oil that Japanese refiners acquire from abroad. Once the objective is understood

the means to reach it are not difficult to envisage: the obvious idea of a counter-OPEC of consumer countries is not necessarily relevant; it would be sufficient for any one of the biggest oil importing countries to make it known that oil imports would not be acceptable if their cost were to exceed a certain level; if the tax levied by producer countries were to rise, making it impracticable to keep within stated limits, short- and medium-term measures would be taken to concentrate on imports from such countries which would prefer to maximize their revenue by increasing the volume of these exports. As long as such policies were pursued judiciously and with moderation they would fit almost painlessly into the existing setup of international oil, without dispossessing the oil companies whose role is, and for some considerable time will be, indispensable. If the thinking of at least one such company, 'leaked' recently, is a relevant indication, the awareness of the situation has reached these quarters even if that concept is being used mainly for tactical purposes.

In fact the mere knowledge that the true nature of the problem has been fully analyzed, that there *are* limits to further exactions, that the world will not be held to ransom indefinitely, will have a sobering influence and the beneficiaries of today's situation will be reluctant to endanger what they have, for the sake of what some people think they could still obtain on top of it in the future.

'As is' has always been a commonsense concept, the current level of benefits should be recognized and an endeavor to force it down deliberately would be out of place. On the other hand, any attempt to push that level up still more is likely to be resisted, as long as such a move was not the result of a genuine change in the supply/demand situation.

Once more: the alternatives open to oil exporters and importers are simple. If trade is free all round and prices are allowed to find their own level – well and good; if, however, one side organizes itself to be able to attack in close formation, the other side cannot afford to keep smiling: the world 'cannot endure permanently half slave and half free.'

14 New Frontiers

This brief piece is included in order to cover an important element of Paul Frankel's career – that of the seminar/conference/study course designed to educate younger oil executives in the economics and politics of the industry. The most important of these was the 'Frankel Seminar' which was started in 1969 and is still run by PEL today. It was designed primarily, but not exclusively, for PEL's own clients; it gathered a wide range of participants from companies and governments and similarly provided a variable menu of speakers from both public and private sectors. Frankel also started in the early 1960s a seminar at Northwestern University, Illinois, and was closely involved with an annual oil industry economics course run by the Institute of Petroleum in London from the 1960s onwards. His role as an educator has been very important to him and, until in more recent times entrepreneurial activity has multiplied opportunities, there used to be few courses available outside the multinational companies' own training programmes and those organised or inspired by Paul Frankel.

There is nothing static about the oil industry which for a long time has been in the habit of doubling its output every eight to ten years but there have been a few periods in which there has not only been a 'rolling readjustment' but a change of direction and it looks as if we are now in one of these periods.

The New Frontiers can be envisaged on different levels of the industry. There is first a spectacular broadening of the search for oil in the geological, technological and geographical senses of the word. The emergence of North and West African reserves and more recently those of the Far North represents the first major breakthrough since the discovery of Middle

Excerpt from the opening statement at the Petroleum Economics and Management Conference organized by the Transportation Center at Northwestern University, Evanston, Illinois, March 1969.

East oil some thirty and more years ago. This shedding of conventional concepts and the search on the strength of fresh theories has opened new vistas.

The progress made, which has greatly enhanced our ideas of potential oil reserves worldwide, is not only due to a New Geology but to the sum total of technical improvements – those optimists were right after all who believed we could rely on the unproven reserves of technology. Progress was not limited to the means of locating oil and natural gas; the ability to tackle offshore drilling in deep water and under awkward conditions, together with an advanced pipeline technique has made oil reserves accessible which were beyond our reach only a decade or so ago.

Apart from the increments in global reserves the oil and natural gas found in the last decade have to some extent reduced the erstwhile semi-monopoly position of the Persian Gulf, although one law of nature has not been contradicted yet: oil in large quantities is usually found in difficult climes and with a long way to go to the markets.

Another New Frontier of the oil industry is provided by the fact that it has become the main supplier of the rapidly growing petrochemical industry. In Great Britain for instance the demand for naphtha is now within reach of the total demand for motor gasoline but the change is not on a quantitative plane only; in fact, the two industries are beginning to coalesce. This is due to a number of reasons including the disposal of certain by-products of petrochemical operations in the mainstream of the oil industry. It is also a safe prediction that oil companies will continue to integrate forward into petrochemicals to share in the profits of that sector but also to secure the outlets for products from their refineries thus for their crude oil. Yet, on the other hand chemical concerns will tend to integrate backwards into oil, vide Union Carbide and Allied Chemicals in the United States, ICI in the U.K. and Badische Anilin in Germany. This they are doing to diversify their risks but also to safeguard their feedstock supplies.

Finally, there appears to be in the offing a major restructuring of retail distribution, especially of gasoline, where the wheel has turned full circle: gasoline started by being dispensed at the kerbside or in the repair shop, then graduated into self-contained service stations but now the service stations, whose costs have risen to an extent no longer justifiable by the sale of motor fuel only, tend to attract in earnest sidelines in the TBA and related fields or in catering. Also gasoline selling points have become part of supermarkets and shopping centres.

15 Topical Problems

Most commentary on and analysis of the oil industry has tended, particularly in more recent years, to cover the upstream, with emphasis on OPEC and consumer/producer/company relationships. Comment on the downstream has often been concentrated on petrol prices and retail competition. This 'topical' is a reminder that Paul Frankel started his career in the market-place and was as interested in downstream markets as in upstream policies; many of his earlier writings during and immediately post-war were concerned with the market and with transportation. As always, the theory and practice of competition are not far from his thoughts.

There is a saying that to evaluate the latest news one should go back to first principles. We believe therefore that it is appropriate for once to consider as 'topical' the background to some of the developments with which we are now faced.

In the news about oil industry matters we find most of the time reports about severe competition in certain markets such as Western Germany and Switzerland but at the same time we see that in other markets, e.g. South Africa and, for a long time, Holland, there prevails competition of a comparatively mild kind. Then again, we have countries such as France since the 1920s and the U.S.A. since the 1930s where the operations of the oil industry are subject to certain governmental controls and restraints. All these phenomena – competition and co-operation of oil companies and governmental controls – are different facets of some basic problems which are common to a number of situations.

Taking Western Germany as a test case for the first variant – intensive price competition – that phenomenon dates back to the establishment of large-scale refineries in the Federal

Republic starting in the early 1950s. They were designed to serve altogether new markets – light and heavy fuel oils – in which there was not, as was the case to some extent for motor fuels, a fairly consistent pattern of market shares of the companies – thus *all* companies were keen on establishing a proper position for themselves. Fuel oils, especially the heavy type, were sold by those who quoted the lowest price and the companies seemed to be anxious to use their new refineries to the greatest possible extent, regardless of current returns – low cost of the last part of the capacity in a plant where fixed costs are high and variable costs are minimal.

Fuel oil demand after World War II growing more quickly than that for motor fuels, the pressure extended to the latter since fuel oils could not be produced without a certain amount of light ends becoming available and it resulted in a virtually unbroken series of price reductions for gasoline. The prices at the pump were also influenced by the emergence of cut-price stations which, apart from imports (being the surplus of Italian refineries), were to a great extent fed by marginal quantities from the main oil companies which they pushed on to the market over and above their own direct (branded) outlets.

From this example one can conclude that extreme price competition is likely to arise in those markets in which rapid development takes place and which are subjected to offers of 'incremental' quantities. In contrast there are markets in Africa or, until the last few years, the U.K. and Holland, which for one reason or the other are not subject to incursions from the outside or to particular stresses within and where the same oil companies behave differently from, say, in Germany. In such circumstances there is a tendency inherent in an oligopolistic situation to show a certain respect for the competitors' market positions and to indulge in alternatives to price competition, for instance in extensive propaganda, investment in facilities, etc. All this is reasonably workable even in the absence of any agreement, written or implied, between the companies involved.

Looking at the overall world situation one can say that, with the increasing ease of communication, the development of free-trade zones, the growth of demand and the general availability of crude oil, the tendency would be towards intensified competition of a more open character. In fact, some of the 'healthy' markets, such as the U.K., Holland and some of the smaller markets in Africa and elsewhere, have softened or are in the process of doing so.

If market control by oil companies has become less prevalent the incidence of governmental controls is probably as great as ever. Such governmental controls are mostly designed either to maintain and safeguard indigenous sources of supply in the face of lower-cost imports, or to back local or national operators against the pressure of international integrated and diversified oil companies.

Experience over the last thirty years or so has shown that companies which do not cover the whole range of operations from production to marketing and companies which operate only in one country or in one region are not capable of standing up to the competition of those companies which are internationally diversified and integrated.

Examples in several oil importing countries, primarily in Italy and Germany, have shown that such non-integrated and local operators as there existed either dropped into insignificance or tended to be drawn into the orbit of international companies in whose hands their assets were worth more than in their independent state. Only where the governments of such countries took a definite stand – which consisted mainly of creating state companies such as ENI in Italy, OMV in Austria and ERAP in France – was it possible to maintain a sector in the oil industry which was not subject to international, i.e. foreign, control. The existence of such 'unsinkable' state-controlled companies was in some countries, especially in France, supplemented by regulations which gave the authorities a certain amount of control over imports, the establishment of refineries, etc. so as to determine to some extent the activities also of the foreign-controlled enterprises. In Spain and Portugal the influence of the government, going as far as the monopoly position of the Spanish CAMPSA, has been all-pervading.

The system which now appears to be evolving in Germany consists of discouraging foreign companies from taking over the remaining German enterprises and also of providing official and other facilities for development of an economically justifiable programme for oil supplies.

On the face of it, and especially if our forecast of increasing price competition in wide-open countries is to the point, the incidence of a governmentally-determined oil industry could result in the consumer paying rather more than he might have had to if competition of all comers was given a free rein. This is due to the fact that quotas generally discourage price competition and that national and local enterprises tend to need a certain price level to cover their sometimes high cost.

The motives for what is called 'intervention' by the governments are to a great extent of a political nature, i.e. they stem from the dissatisfaction with the 'commanding heights' in the most important energy industry being altogether under the control of foreign corporations. There are, however, also some straight economic considerations, which would provide a case against laisser faire.

As we have shown above, there is an alternative to the competitive tendencies in the oil industry and here and there we have had, in the not too distant past, periods of cooperation among oil companies which aimed at price maintenance. Could it perhaps be possible for the international operators, once there are no local competitors, to use their 'collective common sense' one day and to establish a situation in which they could arrange things to their liking? If these companies were ever so minded, the absence of any firsthand knowledge, which only nationally organised companies could provide for countries where no international oil companies are domiciled, would make it difficult for their governments even to know what had hit them.

When it comes to the question of governmental control the position of the main oil-exporting countries – since 1960 organised in OPEC – has to be taken into account. Such control takes place on two levels: since in virtually all countries (except in most of the U.S.A.) the subsoil resources belong to the government there is a prima facie tendency towards control especially now that the concept of over-riding sovereignty has invalidated all ideas of an inviolable contractual relationship of the (mostly foreign) operator to the concession-granting authority. The remedy for such control lies in the competition (actual or potential) between the governments of oil-bearing countries bent on attracting risk-taking investors and on keeping their exports as high as possible. If such competition should be effectively inhibited and should the perennial dream of the OPEC Secretariat to establish quotas of output (international proration) materialise in one form or the other, there would be established a degree of control which would be bound to change the balance of power between producers and consumers.

In view of the possibility that the international oil companies might see fit to take the line of least resistance and go along with the OPEC-type policies, one should not be surprised if oil-importing countries were to find that the need for nationally-orientated oil companies and for well-defined oil policies of their own were to take on a new and pressing character.

16   Oil Prices: Causes and Prospects

This piece was written for Platt's *in January 1971, just before the Tehran Agreement was signed (on 14 February 1971). It is illuminating to read it now, particularly the latter part of the article, in the light of all that has occurred since. Frankel correctly noted the changed psychology of supply/demand perception as an 'absence of surplus'; this was to remain the overriding assumption of all policy-makers until the Iranian Revolution and the second price crisis of 1979–80. His side-swipe at the companies whom he suggested might benefit from hardening of price as much as the OPEC producers was, perhaps, fair but the implication of possible collusion between the two much less so. What, however, is more interesting in retrospect is Frankel's pre-vision of what later would be called recycling and of the plight of the LDCs; also the absence of countervailing power amongst the oil consumers. 'The answer', he suggests, is that 'enterprises – privately or governmentally owned – can successfully operate only within an international framework of rights and obligations from which could emerge a world price level with which both sellers and buyers alike can be expected to live.' He did not, in 1971, foresee that this might be the $18 that so many assiduously propose in 1988 but he could justifiably feel a glow of satisfaction, even if the chances of creating a formal international framework today are no higher than they were seventeen years ago.*

I

The events of the last few months, which have led to a dramatic reversal of a downward trend of oil prices of some fifteen years' standing, can be appreciated only if we see them in their proper political and economic context.

As the political situation in and around Arab lands looms large in oil affairs, and since governments and the prerogatives of sovereignty play such a substantial role, we tend to overlook the fact that political power as such can in the long run shape trade effectively only if it is in line with, or at least buttressed by, the underlying economic factors involved. In our case now, the actions of a regime such as the one in Libya could realise its ambitions with such apparent ease only because, by design or luck, it hit at a point of least resistance.

All this cannot be understood if one does not recall the remarkable if little noticed fact that there co-exists a very wide range of costs of crude oils and, beyond oil, of sources of energy at large.

Although traditionally U.S.A.-produced oil was cheap and plentiful, both features had ceased to obtain by the end of World War II and the emergence of prolific and low-cost crude oil in the Middle East happened to coincide with a 'Fortress America' domestic price structure. In a pamphlet published several years ago, I have traced the remarkable story of the means by which the new low-cost crude oils were 'eased' into a market still, if to a diminishing degree, deter- mined by higher-cost crudes: it is this wide range of crude oils where the actual cost of the lowest ones was but, say, a tenth of the highest, with the latter being yet kept in the picture, which provides the background of the current situation.

From this 'economic rent', i.e. from the advantage of the low-cost producer, were derived the huge profits which the few privileged oil companies then involved made in the first instance and – from the early fifties onwards – the tax take of the oil producer countries. If this 'rent' had not been there for the asking – with the connivance of the producers of U.S. oil and of European coal, who preferred higher oil prices to the lower ones which would have accelerated their own demise – the remarkable and probably unprecedented fact would never have materialised that an export levy up to ten times the actual cost of producing the crude oil could be imposed on it.

Whereas in the course of the last ten/fifteen years the oil companies have shared in this 'rent', their part of it has – some pockets of high margins notwithstanding – been gradu- ally whittled down by the due process of competition, that part which came to be allocated to the producer countries has tended to grow per barrel and, in view of the continuing rise in exports, still more so in terms of wealth transfer from oil importing to oil exporting countries.

In spite of the successful defence of the per-barrel take of the countries – the great achievement of OPEC – in a world in which oil prices went down almost continuously for more than ten years, some observers were to my mind justified in expecting that once oil exporting countries were themselves making decisions as to the level of their production – a task hitherto mainly performed by their foreign concessionnaires – they would be faced with the age-old dilemma of any operator. The dilemma to which extent he should optimise his revenue by stressing the volume, even at the expense of some margin per unit, or by concentrating on the margin even at the expense of the last part of his potential output. The assumption, which I shared until recently, was that once the decisions were in fact taken by governments (or by their national companies) the latter would behave exactly as oil companies have done for most of the time: they would realise that it pays to sacrifice some part of the margin on some part of their output provided that (as it so often does at least in the first instance) such a move substantially increases the quantities sold with the ensuing result of a rise in total receipts. If one makes many times one's cost by way of 'take', and can produce additional barrels at constant or even reducing cost, it is almost inevitable that one should try to broaden the basis of such margin.

The knowledge that, what can be described as 'economic gas pressure' which pushes additional low-cost barrels towards the market place, might result in a progressive lowering of the price level, prompted the perennial endeavours in OPEC to establish some sort of programming of production alias international proration alias worldwide quota cartel. Significantly Venezuela – a low-reserve and high-cost country – was the protagonist of a concept of a fully-fledged production programme and equally significantly nothing much ever came of it: in the process of maintaining or enhancing the 'take' per barrel OPEC members are allies but when it comes to the volume of exports they are inevitably each other's straight competitors.

II

The cataclysmic events of 1970 reversed the trends of the last decade or so and it is for us to analyse whether what has led to them is fortuitous and ephemeral or whether we have witnessed yet another turning point.

The immediate circumstances were to be found in the sphere of transportation: just when the demand for additional tanker space caused by the closure of the Suez Canal had been met by more and bigger tankers (well placed for the Cape route) and, true enough, by a spectacular shift of supplies for Europe to nearby African loading points, and thus freight rates had about reverted to a 'normal' level, there occurred the closure of TAPline in May 1970 and it also became obvious that all previous estimates of oil demand had proved to be much too low, especially at the heavy end of the barrel. This was due not only to higher demand for energy but also to coal in Europe fading out rather more quickly than had been assumed, so that oil had to take the full brunt. Looking further ahead, it is also due to atomic energy falling once again into one of its recurrent troughs of delays and disenchantment.

This convergence of unrelated circumstances made it possible for the Libyan authorities to land a first-rate coup: they made excellent use of the shortage of means of transport, which rendered particularly valuable Mediterranean crude which, destination Rotterdam, needed only about 30% tanker capacity and destination Marseilles about 15% as compared with oil to be hauled from the Persian Gulf. In exploiting this particular chance they hit on a device the use of which was directly opposed to the more likely concept of optimising by volume: they deliberately reduced exports by introducing 'Allowables', a term hitherto unknown in the Eastern Hemisphere. Although they did cut down their current income, they might in the end be justified since they managed to start an upward movement of the oil producer country 'take' whose end we are yet to see. It is this extension to suppliers other than in the Mediterranean, who do not share its particular logistic advantage, that indicates the real nature of what is going on.

The clearly distinguishable background to these developments is provided by the fact that not only, as I recalled earlier on, have Middle East (and in a different way African) crude oils lower cost than other sources of supply, hence the 'economic' rent available, but that *all* oils at their early 1970 prices were to their buyers in most markets considerably cheaper than any alternative source of energy – coal, atomic energy, hydropower and whatever; furthermore, oil has been the only energy source which can be relied upon to meet the greater part of current demand and for that matter all of incremental demand – at least for most of the seventies.

Hence there is another 'economic rent' available and in this sense those who have maintained that oil was being sold too cheaply had something to back up their arguments when they said that it was intra-oil competition which frustrated the otherwise attainable goal to sell it in each case at just below the price of alternative sources of energy, which would have been the classical procedure of a monopolist.

In fact the crude oil prices as practised until recently represented a three-way split of the economic rent available: (i) a substantial slice went to the producer-country – up to and sometimes more than $1 per barrel; (ii) a sizeable but over time shrinking slice – until recently, say, 20 to 40 cents per barrel – went to the oil companies and (iii) the remaining considerable part – difficult to boil down to a number or even to a range – was enjoyed by the oil consumers. It is this 'consumer rent' which is now under pressure.[1]

Surely if the oil industry had been less competitively organised and/or if OPEC had previously been able to set up an effective proration scheme, the general oil price level could have been higher than it has been so far: now the accidental coincidence of logistic bottlenecks with unexpected increase in demand has in a way called the bluff and the success of the Libyan tour de force has lit up a scene hitherto shrouded in a kind of fog made up of historical precedents. The events of the

---

[1] It is obvious that it is mainly this rent which made it possible to levy on petroleum products such very high excise taxes as have been imposed in most countries. Yet the incessantly repeated comparison of these taxes with those payable on crude oil to the producer country is arrant nonsense: it has been pointed out before that these excise taxes are just part of a fiscal system *within a country*, whereas the export levies are borne by the buyer in *another country*.

summer of 1970 in Libya have something in common with the famous Paris 'evenements' of May 1968: fortuitous in themselves they disclosed the basic weakness of a prevailing setup.

There remains to analyse what, if any, the forces are which could stand in the way of a further progress of shifting the shares in the 'rent' still further towards producer interests. In this respect the defensive position of the oil companies is rather weak: they are prisoners of their investment and as long as they retain a modicum of cash flow – or at least hope to do so – they will, under pressure, tend to give in to the demands of the sovereign power which can cut it off: more than two years ago I had suggested that instead of 'host country' one should talk of 'hostage country'. In fact, considering the role of oil companies as purchasing agents, as it were, of oil importing countries one may well ask what would happen if it was left to the kidnap victim himself to negotiate the amount of the ransom – a ransom to be paid actually by a third party. The third parties are the consumers or consumer countries to whom in the circumstances the incremental tax cost is now being passed on.

The position of the international oil companies has changed in yet another respect: their global operations used to imply that they had a built-in supply flexibility which strengthened their hand vis-à-vis any one host country and provided an assurance of security for their customers. Indeed the failure of nationalisation in Iran was altogether due to the companies being able to isolate Dr. Mossadek by drawing supplies from alternative sources willing, nay anxious, to make available low-cost oil.

Yet less than twenty years later there would be in such circumstances no spare capacity (even if OPEC had not provided a withholding procedure for such and similar cases): the official demise of the U.S., and, for that matter, of Venezuela, as emergency oil supplier to the world has finally swept away whatever illusions may have lingered on of late.

III

What does all this mean in terms of oil prices in the short, medium and long term? There is no doubt that we are entering a period of higher crude oil and of still higher products prices – especially for distillate and residual oils.

This is obviously due to the greater tax burden on crude oil, to higher freights – often over longer distances – and, downstream, also to the absence of surplus refining capacity in several areas.

There is comparatively little resistance among consumers and by most consumer-country governments against price increases, simply because the tight tanker situation has engendered a worry about potential shortages which will escalate if this winter turns out to be cold: obviously everybody is concerned about higher prices but the fear of cold homes and of idle factories is much more immediate and telling. The apparent success of the unprecedentedly heavy propaganda barrage the oil companies have laid down to persuade the consumers of the inevitability of massive price rises will, incidentally, tend to make it more difficult for them to withstand the pressures from the OPEC side than it would have been anyway.

When we analyse the Higher Cost claims of oil companies we come across yet another 'economic rent': operators with large tanker fleets – most of it built or chartered at lower costs than are prevailing now – obviously have an average cost considerably below the high of the range applicable to current charters and to ships ordered now.

The important point, however, is what is likely to happen once transportation costs have come down, following the coming into the market of more tankers and the completion of all or at least of some pipelines now in the project stage – to say nothing of the reopening of the Suez Canal. If I have read aright the portents mentioned earlier on, oil prices will then still remain considerably higher than they were until some months ago.

The oil companies whose profitability had been well below par in some areas such as Germany or Italy, will endeavour to widen their margins whilst the going is good and some relaxation of competitive pressures might make it possible to hold on to such gains for some time to come. Consumer countries, especially in Europe, may take a somewhat lenient view of such procedures, partly because downstream profits are taxable in their own countries, but mainly because what matters more to them is the foreign exchange burden which massive energy imports involve.

Once the transportation bottleneck has been duly bypassed there will be no question of oil shortage – except in the case of straight political impediments – but what will distinguish the position from that prevailing until 1970 is the conceptual *absence of a surplus*. This will firstly be due to the higher level of demand as M. Abdesselam, the Algerian Minister of Industry and Energy said recently in Caracas: 'We have now reached a point when in contrast with the course of events during the past decade, demand for energy in the major consuming centres, contrary to all statistical forecasts, exceeds supply by a margin which has never before been equalled'. This statement is quoted here, not because it is correct, tanker shortage apart it isn't, but because it shows the frame of mind of those now in command.

They have seen how rich the rewards can be of pressing on with the gobbling up of ever larger slices of the cake, i.e. of the famous economic rent available; since they can do this without endangering their precious unity, it might be quite a long time before they discover once again (as they probably will eventually) the alternative charms of optimising by increasing the volume of sales in competition with others.

The reason why I believe this to be a likely course of events lies in the fact that the, deliberately chosen, 'they' fits admirably not only the OPEC members but in fact the oil companies themselves. To all those who *have* reserves and established downstream positions the prospect of their enhanced valorisation by a – perhaps temporary – reduction of what I have called 'economic gas pressure' should be rather attractive.

The difference between the situation in the late fifties and the one now unfolding lies to some extent in the fact that, with a very few exceptions, the then 'newcomers' have failed to establish themselves properly: this was partly due to most of them having failed to 'strike it rich' and also to having been hamstrung by uneconomical concessionary, or as one now says contractual, terms; and those who did hit jackpots of sorts have been subjected to a degree of harassment which they could face with less ease than could the, by then, well-established and diversified operators of the preceding generation; the meagre results of the bulk of American newcomers and the but limited success in the field of exploration of organisations like ENI and ERAP have sapped the incentives for like-minded 'third generation' operators, hence it is, in the generally unpromising climate prevailing, unlikely that the consolidated fronts should be disturbed soon by further intruders. Without the irritating stimulus of the gadfly the sitting tenants are likely to revert to their erstwhile oligopolistic stance.

Is there thus no corrective antidote to higher prices? Precious little of it can be seen just now at least in the isolated context of oil industry bargaining. It is therefore necessary to look at the problem against the wider background of world affairs.

IV

When considering the repercussions of crude oil cost increases on the oil importing countries we must distinguish between developed and developing countries.

The increase in the price of energy will tend to push up the cost of manufacturing processes, especially of those for which energy is critical: it will also tend to accelerate the fall in the value of money and it will certainly reduce the overall buoyancy of industrialised countries which have contributed so prominently to the world-wide growth phenomenon of the last twenty-five years. It must be remembered, though, that such increase of cost, if it affects *all* the industrialised countries about equally, has no discriminatory quality, i.e. the main contestants, being similarly affected, keep their respective ranks in the competitive setup, admittedly with a certain bias in favour of the U.S.A., U.S.S.R., Canada and perhaps in the future of Australia, for whom energy imports are in varying degrees supplementary.

The increase of foreign exchange burdens could be more critical, for countries whose position in that respect is uneasy enough as it is: whereas the tax increases by OPEC look paltry compared with the consumer prices for petroleum products (including excise taxes) they represent a considerable added burden on the level of import parities.

There is, however, one facet which looks like alleviating this additional burden in the case of some of the oil importing countries. Much of the oil exporting countries' 'take' does in fact return in one form or the other to where it came from in the first instance: if the money (and especially the additionally earned part of it) is not immediately earmarked for local projects, it flows back into the pool of international investments which, for the time being inevitably means to the U.S.A. and Western Europe. If, however, it is used in the process of industrialisation or even in purely hedonistic pursuits – and that includes the vast purchases of armaments – it will to a great extent take on the form of purchases from industrialised countries. Under all headings the U.S.A. is least affected and even somewhat favoured, whereas the U.K. and France will be hit but they both have sizeable remedial advantages – including the profits of 'their' oil companies.

The position is radically different in respect of underdeveloped countries, some of whose oil imports represent their main foreign currency outgoings. Is it really acceptable to see such a large part of the 'economic rent' being used to pile up yet more wealth in the few fortunate countries, some of them exceedingly small, where oil happened to be discovered, when so significant a part of this great and growing flow of money comes from a welter of particularly poor countries? As it is now, India and Pakistan subsidise, say, Kuwait and Abu Dhabi, whereas, say, Brazil and Jamaica subsidise Venezuela.

If such increase in oil prices was not engineered by governments and was not based on one form or the other of deliberate restraint of the flow of oil, one could let the chips lie where they fall – however, as things are today and as they may develop tomorrow, one should not be surprised to see underdeveloped oil importers raise this matter at whatever international forum they might consider to be appropriate. This leads to the question whether and to what extent the world at large will acquiesce in a situation in which a minute part of the world's population strives to cash in to the full on a monopoly, both fortuitous and contrived.

True enough, in the longer run, the more such a monopoly

position should be abused the quicker would the endeavour to bypass it lead to tangible results: the higher-priced and the less secure the imported energy should become, the greater would be the incentive to diversify by multiplying the sources of supply, by fostering exploration for and development of indigenous reserves of oil and natural gas and by bringing forward atomic energy, gasified coal, tar sands, shale oil and all. Most of these endeavours will take up vast capital resources which, in more equitable circumstances might be used for tasks other than that of the substitution of a source of energy which involves an incomparably low actual effort.

In the shorter run, if my interpretation is to the point, there is but little countervailing power in the hands of industrialised consumer countries and I do not therefore believe that, say, Western European governments would be better placed than oil companies when it comes to straight price bargaining with producer countries – they might at this juncture even do worse than do the oil companies, the latter's recently dismal record notwithstanding.

Yet, when looking at the sum total of the relationship of the industrialised oil importers to most of the OPEC countries, it emerges that the latter have a certain range of needs: sometimes access to markets for produce other than oil and in virtually all cases access to the capital markets of the world.

This brings us right to the centre of the problem: the events of the last few years, stemming from the doctrine that a sovereign state can override all commitments it has entered into, have downgraded international intercourse to the level of the jungle. Hence the mounting difficulty of attracting capital into double-risk ventures: double-risk because superimposed on the technological risk of failure in exploration, with which the oil industry has always been able to cope, is now the political risk of the agreed terms being altered against you, should you have the very success whose expectation was the only reason for taking the technological risk in the first place. Who will buy a lottery-ticket if the winner of the jackpot is subjected to iron-clad arguments and if – lest he be altogether dispossessed – he has to make a foul compromise with the lottery holder?

What is needed – especially for the considerable amounts

required on all levels to meet energy demand by the end of this decade – is a new kind of status of international investment and trade where each party acts as a provider of what it can best contribute but can have the reasonable assurance that the rules of the game will not be changed at half-time.

We might in fact have reached the point where the traditional idea that commercial and industrial relations are best left to the operators who act as 'buffer' between governments is no longer valid. The answer in matters oil might be that enterprises – privately or governmentally owned – can successfully operate only within an international framework of rights and obligations from which could emerge a world price level with which both sellers and buyers alike can be expected to live.

V

Although I am here for an analysis of trends rather than an endeavour to assess Stop Press news, the initiative taken mid-January by the oil companies in response to the challenge of OPEC as a whole and to that of the North African countries in particular is of such symptomatic interest that some of its aspects must be brought into focus:

(1) Whereas previously the function of OPEC as a negotiating instrument had been discouraged by the oil companies, now 'collective bargaining' on both sides is being sought as being the only means of stemming the tide of high-handed action of some governments.

(2) The acknowledgment of OPEC involves the shedding of the antitrust fiction as far as oil companies are concerned: those facing a monopoly have to hang together, lest they hang separately.

(3) The proposal of a grand design for prices and their adjustment should or at least might make it possible for the conservative members of OPEC, whose goal is in the realm of economics, to prevail upon those who are in the first instance bent on power.

(4) The terms of such a settlement would inevitably be very onerous in terms of oil prices; they might become accept-

able to consumers if they re-established a kind of consistent and reasonably constant framework of rights and obligations.

(5) Whereas the competitive working of the market place could be considered as being self-regulating without the need to involve governmental prerogative, a masterplan worked out between OPEC and the oil companies would require to be backed by consumer country governments who have to foot the bill.

The fact that taxes imposed by oil producers are in fact paid by oil consumers calls for their being consulted all the way, as is normally the case when Commodity Agreements are being negotiated. 'No taxation without representation' is a time-honoured principle.

(6) An 'armistice' in the contest which has been described earlier on will not save all concerned from the need for a gradual but fundamental approach to the real problems of co-existence; radical measures taken in certain quarters for the sake of Instant Power would tend to act as a catalyst and would lead to a rapid and comprehensive closing of the ranks with results which might be irreversible.

17   Note on Current Oil Policies

Following the revolutionary changes imposed on the oil industry by the events of 1970–71 – the Libyan and Tehran price agreements followed by the Geneva currency agreement and the General Agreement on Participation – Frankel was convinced that some form of 'countervailing power', as mentioned in the previous Platt's *article, was required. He would later push this strongly – in the context of post-1973 developments and the IEA – but in 1972 he was already working on ideas that derived from his own thinking and from the consultancy contacts he had established with the non-major companies in Italy, Belgium, West Germany, Austria and Spain. This is an unpublished note describing the objectives of what was known as the Zurich Group. It consisted of ENI, Veba and ÖMV; it was so named simply because meetings took place in Zurich. In the event, the initiative produced no specific outcome but it was a useful mechanism for the spread of information and of preliminary analytical thinking which no doubt had its beneficial effects when the 1973 price crisis triggered the creation of the IEA. The 'French initiative' referred to was the government-inspired settlement of CFP with the Iraqi government in June 1972 over the nationalization of IPC; the other IPC partners only reached a final settlement nearly a year later.*

I

The situation as it prevails at the moment is determined by the following elements:

(1) The eight international oil companies have lost much of the arbitrary power they have had, due to:

(a) the fact that it is now generally assumed that the period has come to an end in which the potential productive capacity of Venezuela and the Middle East appeared

Unpublished note describing the objectives of the Zurich Group, September 1972.

to exceed greatly potential demand for a long time to come, and

(b) the fact that the resulting increased power of the supply side has in the first instance been to the benefit not of the oil companies but to the by now effectively organised oil producing countries (OPEC).

(2) On the other hand the oil companies have more recently seen their position strengthened because:

(a) the producing countries have realised that the oil companies' rôle in respect of investment planning, finance and of transport/refining/marketing is still essential;

(b) some OPEC members fear that without the cohesion provided by the oil industry the competition between the producer countries themselves would be greater and thus their returns lower;

(c) the political power of the U.S.A. has re-emerged as a result of the successes so far of the Nixon/Kissinger policies. This has undoubtedly influenced the Shah towards the extension and deepening of the Consortium agreements and is likely to make a deal also with Saudi Arabia possible.

(3) The availability of 'nationalised' and of 'participation' oil of OPEC countries (and the 20% of production which NIOC will have at its disposal at cost plus a moderate amount) will in due course lead to a loosening up of the major company control but it will take time and the direction of change is not yet clear.

II

The problems of oil importing countries (especially of those which do not have a substantial share in the existing international oil industry) are roughly as follows:

(a) they have to face further (probably substantial) price increases which the OPEC countries will be able to impose, simply because there will be no effective oil surplus; also, at least some of the countries will tend to restrict production because they will expect to get higher returns per barrel later on.

(b) in the absence of surpluses the oil companies themselves might not indulge in price competition and will try to increase their margin; thus those consumer countries which do not share in the company margins will be doubly affected.

The overriding interest of consumer countries lies in the investment to develop the Middle East potential being made now and in the future. The making of this investment is more pressing and vital for the consumers than it is for either producer countries as a whole or for the existing oil companies: the last two would also benefit from a state of shortage because the reserves which they control would then become more profitable, whereas the consumers would be hit by a real shortage *and* by sky-rocketing prices.

III

In the circumstances the overriding interests of the consuming countries are:

(a) to make sure that the investment necessary to find more oil (and natural gas) and to develop known reserves – especially in the Middle East – is forthcoming; this to be achieved inter alia by their themselves providing the necessary funds.
(b) to safeguard their vital interests through the policy just mentioned by:

    (aa) avoiding a genuine shortage of hydrocarbons world-wide;
    (bb) making it more likely that competition for access to markets between producing countries and, in another way, between oil companies should be maintained.

Whereas the policy objectives just mentioned are identical for all consumer countries the position is different in respect of (b) (bb) above:
Whereas the U.S.A. and the U.K. (and to a lesser extent France) would benefit to some extent from a tightening of the market and the resulting absence of competition, those countries like Japan, Germany, Italy and Spain which do not have the compensating advantages of benefiting from better margins would be hit much harder.

Consequently their policy must be directed towards:

(a) seeing to it that the spirit of competition for markets does not disappear altogether (this partly to be achieved by the investments mentioned at the beginning of this Section III); (b) seeing to it that, especially if the endeavours just mentioned under (a) are not, or not altogether, successful, *their* own companies at least participate in the profits made in such non-competitive situation. Also they would gain a high degree of knowledge of what is going on and obtain at least some influence on planning (including the critical investment decisions).

IV

Although the endeavours of national companies such as ENI, ERAP, Deminex, Hispanoil and those of Japanese interests to find new oil reserves have yielded some interesting (although sporadic) results, the main part of world oil reserves remains in the hands of the eight international companies, and means have to be sought to make the investment potential of other consumer countries available for the purpose of development and to give these countries a share in the decision making.

It is likely that a concerted strategy of countries such as Japan, Germany and Italy would be more effective vis-à-vis oil companies and producer countries than could be the isolated manoeuvres of any one of them. Also competition among them for access to sources of supply would weaken the position of every one of them and thus drive up the 'price' they will have to pay. On the other hand it has proved difficult for any one of these countries to start cooperation because the others are inclined to prefer not to join a grouping initiated and therefore led by another country.

All these considerations have, at the beginning of 1972, led to the formation of a Study Group consisting in the first instance of Deminex, Hispanoil, Japan Petroleum Development Corporation and Oesterreichische Mineraloelverwaltung A.G., which has since become known as the 'Zurich Group'. Its aims are roughly:

(1) To provide a means of mutual consultation of its members on all matters of mutual interest.
(2) To develop plans for joint approach to the problem of acquiring positions in rights to existing reserves.

These objects are to be pursued on two levels:

(a) Where events lead to a fundamental restructuring of the existing setup Group members might join new 'consortia' to work alongside with the traditional concession holding companies or to succeed them by mutual agreement.

(b) Should it not be possible to realise the approach (a), there might be pursued jointly the possibility of coming to terms directly and without reference to the present major companies with the national companies of the producer countries. Although the latter will want to establish their prerogatives there will be a definite need for a sustained effort coming from abroad in respect of long-range investment planning, actual investment and coordination with downstream activities – transportation, refining and marketing. These indispensable activities could (and will inevitably) be carried on by oil companies in a form to be determined and to be named, under the cover of 'purchase' contracts. For such operations comparative newcomers, such as is the type of Zurich Group members, appear to be particularly suitable.

The scope for level (a) strategy and tactics appears now to exist within the framework of the 'French initiative' in respect of Iraq. The level (b) problems will have to be determined in the light of the Iraq outcome and will depend on the setup established for the first phase of OPEC country 'participation'.

18   Note on the Repercussions of Crude Oil Price Increases on Developing Countries

This unpublished note is included only as an indication of Frankel's genuine concern for the plight of LDCs over the increasing cost of their oil imports. This was, of course, to become an even more acute problem after the massive oil price increases of October and December 1973.

(1) Since 1970 crude oil prices in sales to third parties at the Persian Gulf have increased from about $1.20/1.30 per barrel to $2.00 to $2.20; it is not unlikely that this steep upward trend will continue and just now it is not possible to say at which level it can be expected to stop.

(2) The dimensions of the problem are:

(a) The actual cost of production of the principal crude oils in the Persian Gulf is of the order of U.S. $0.06 to $0.15.

(b) The total 'tax take' of the Persian Gulf oil exporting countries was until about two years ago around $0.90 and is now $1.50 per barrel. To this is to be added additional government income from the latest 'participation' agreement amounting to around $0.10 per barrel for the principal crude oils.

(c) The profit margin of the operating oil companies (not taking account of marginal operators with higher cost and lower margins) which two years ago amounted to roughly $0.20/$0.25 can now be assumed to be about $0.30 to $0.35 per barrel.

(3) Para (2), above, shows clearly:

(a) The enormous and steeply rising difference between cost and returns which results in 'tax take' reaching levels of up to 20 times the cost of production.

Unpublished note, January 1973.

(b) Whereas the operating oil companies have maintained their margin and may possibly manage to increase it somewhat, it now represents a significantly smaller percentage of the price which determines the requirements for working capital.

(4) The oil tax income of the OPEC countries in 1970 was $7,600 million or $0.95 per barrel. It is estimated to reach in 1975 over $22,000 million or almost $2.00 per barrel. It is likely to be much higher towards the end of this decade.

(5) The figures given in the preceding paragraphs show:

(a) An unprecedented cost/price ratio mainly for the benefit of governments controlling the resource.
(b) An equally unprecedented agglomeration of wealth in a very small region of the world.

(6) The oil tax income of the OPEC countries represents to a large extent a transfer of wealth from developed countries, whose level of industrialisation and of living standards makes them the main users of energy, to the comparatively less developed oil producing and exporting countries.[1] Some of the latter have by now, however, by virtue of their small population and the magnitude of their incomes reached a level of GNP which is considerably higher than that of almost all developed countries; e.g. Kuwait, Libya, Abu Dhabi/Dubai with GNP per capita in the range of $2,000–4,000 compared with U.S.A.'s $5,000, Japan's $2,000 and the EEC's of the order of $2,000–3,000.

(7) The developed oil importing countries' foreign exchange position is undoubtedly affected by the rising outgoings per barrel of oil but there are for them some alleviating factors:

(a) The increased spending power of oil exporting countries results in a rise of their imports (armaments, industrial equipment, services and consumer goods) which originate in industrialised countries.
(b) much of the income not required for uses as per (a) above, does inevitably flow back into the world financial

---

[1] Developed countries have been taken as those with a GNP per capita above $1,000.

system by way of short-, medium- and long-term place-
ment and/or investment.

(8) Developing oil importing countries which in 1970/1971
took about 17% of the exports of OPEC countries now
bear the full brunt of price increases without the alleviating elements mentioned in para (7) above; obviously their
mostly precarious foreign exchange position and the repercussions which a curtailment of energy use would have
on their future development prospects makes them particularly vulnerable.

(9) In the circumstances the problem arises to find a way to
alleviate to some extent the hardship which befalls especially the oil importing countries in the stage of development without prejudice to the basic position of the oil
exporting countries. The justification for trying to reduce
the impact of higher oil import bills lies in:

(a) the degree of net benefit derived from oil export by a
few countries.

(b) the speed of increase in these benefits which leaves no
time for importing countries to adjust themselves to these
changes.

(c) the reasonable expectation that developing countries
which have had the singular luck of having a vital commodity under their sovereign control and are thus rapidly
approaching a level of concentrated wealth would be
guided by a feeling of solidarity with other countries in the
developing stage which have not been so lucky and are
hurt by the very advantages which the oil exporters are
now enjoying.

(10) It has been suggested that developing countries might be
charged lower prices than others, a reduction to be made
possible by lesser tax amounts being due for supplies
going to them; this idea has been criticised on the grounds
that a two-tier price system would be difficult to administer and would give rise to undesirable practices. Thus it
would perhaps be more appropriate to channel a certain
proportion of the 'tax take' relating to exports of oil to
specified developing countries into a fund out of which the
respective amounts would accrue to the importing coun-

try in proportions to their intake of oil; these amounts could be earmarked for certain purposes (possibly related to energy projects). it could also be envisaged that alongside the oil exporting countries the oil companies involved would pay into the same fund amounts according to the ratio of 'tax take' to company profit margins.

There is undoubtedly a need for a study of the ways and means of such an operation but if the principle – i.e. the need to remedy a barely tolerable situation – is accepted, the methods how relief can be provided where it is needed, could be devised without too much difficulty.

19   The Oil Industry and Professor Adelman: A Personal View

For years Paul Frankel and Morry Adelman, Professor of Economics at MIT, had been friendly rivals as analysts of oil and the oil industry. They were rivals only in the sense that there were so few oil analysts writing in the period before 1960, or even 1970. This piece takes as its starting-point the publication of Adelman's book The World Petroleum Market *(The Johns Hopkins University Press, Baltimore and London, 1972).*

In the course of the last ten or fifteen years M. A. Adelman, Professor of Economics at the Massachusetts Institute of Technology (MIT), has become well known all over the world simply by publishing analytical papers on oil industry matters: what he had to say and the way he said it have assured him the attention of industry and government circles worldwide.

The recent appearance of a major work of his, *The World Petroleum Market* (The Johns Hopkins University Press, Baltimore and London) and of an article in the American journal *Foreign Policy* (Winter issue 1972/3) under the title 'Is the Oil Shortage Real?' provides a welcome opportunity for a critical appraisal of Professor Adelman's theories as applied by him to the economic and political problems facing us today.

Before endeavouring to review Adelman's approach I should say that, our durable personal friendship notwithstanding, I have found myself more often than not in disagreement with his theories and thus with the conclusions he draws from them: this is most likely due mainly to our having set out from opposite starting points: he, a learned economist looking with a sharp eye into oil from the outside, I, an oilman, applying my intrinsic understanding of how the industry works to the sphere of economic generalizations.

This is not the place to provide an elaborate critique of Adelman's substantial book – in spite of its length and its somewhat diffuse structure it is recommended reading for anyone who wants to see oil problems in perspective – and I will concentrate on the fundamental differences I have with its author, from which flow almost inevitably the divergencies between his and the oil industry's appraisal of current affairs.

My own assessment of oil problems is based on the characteristic features of a highly capital-intensive industry, an assessment spelled out in my *Essentials of Petroleum*, first published in 1946, and in a number of later publications of mine. Throughout I have traced the all-pervading repercussions of the extreme relation of fixed and variable costs which in inverse ways determine the operators' actions and reactions: on the one hand the rapidly falling cost curve – which moving towards the tail end of the available capacity the marginal costs are low compared with those of the baseload – encourages intensive competitive behaviour; on the other, and for that selfsame reason, there is a deepseated tendency to avoid, or at least to mitigate, the results of this objective situation not only by horizontal concentration and vertical integration, but also by co-operative endeavours leading to varying degrees of 'understanding' among the competitors – some of them engendered by governmental authorities.

Whereas I thus come to the result that the oil industry (mainly due to its built-in anarchic features) is not self-adjusting and therefore tends to generate adjusting measures, Professor Adelman takes the opposite view.

He has on many occasions countered that the insistence on the cost curves' downturn as one moves towards full (or rather optimal) use of capacity did not make sense for crude oil production. Adelman's strictures against 'the mistaken but influential idea that oil and gas production is an activity with increasing returns or decreasing cost as more oil is produced' are repeated in his recent book. He said, rightly, that an oil field (and therefore all fields together) had not diminishing but increasing cost, going up with the progressive exhaustion of the field and the consequently rising unit cost increased also by the additional effort (repressuring and all) called for in the process.

This picture, though correct in its own context, appears to me to be less relevant than Adelman makes out, and that for three reasons: (i) because the phenomenon of decreasing cost is not confined to the production stage but applies equally to the refining, transportation and marketing phases with which Adelman does not seem to concern himself particularly; (ii) because once the field is delineated and developed the cost of the incremental barrel produced is low compared with the original effort and remains so for some considerable period; and (iii) what is true of each existing field, due to its eventual exhaustion, need not be true of the industry where the field's *discovery* does not progress methodically from the easier (and thus cheaper) to the more difficult (and thus costlier) operation in view of the aleatory character much of the exploration activities still have in spite of our now much greater degree of scientific and technological sophistication.

Obviously the full appraisal of the volume potential of the Middle East reserves, and of a cost level much lower than anything previously envisaged, led to a downward turn of what could previously be considered to be the prevailing level of prices. It was the constitution of the oil industry, and its political environment, which inhibited the fuller and prompter spread of these lower costs.

The strength *and* the weakness of the Adelman approach lie in his deepseated belief, correct in its proper context, that it is the relationship of underlying supply and demand which determines the behaviour of the operators and that the price formation which results directly from it remains the Moment of Truth, all endeavours notwithstanding to deform this truth or to delay its becoming effective. It is significant that Adelman has not written a book about the oil *industry* but about the petroleum *market*.

The validity of his argument was to be tested many times over, particularly clearly in the period beginning in the mid-1950s, when the full impact of the (then apparently inexhaustible) reserves in the Middle East and their costs, which were but a fraction of those incurred in the other oil regions, which had hitherto determined the world market, became more apparent. In the fifteen years or so until 1970 the price of

crude oil in fact fell inexorably if slowly – still more than the
dollar quotations showed in view of the decline in its purchasing power. At the time I joined Adelman in the belief that the
downward trend was likely to continue even if I did not go as
far as he did when he thought fit to foretell an fob Persian Gulf
price of $1 a barrel – highly numerate economists sometimes
cannot resist the temptation to achieve what I once termed
'false accuracy'.

Optimizing by Volume and by Margin

Yet, although later on we both proved wrong, when it came to
the sharp upturn of prices from 1970 onwards, Adelman's
thinking was well founded and it is even now worth tracing
the way of his reasoning: so wide a gap between cost and a
going price '20 times longrun incremental cost' – as he called
it in a very recent letter to the Editor of *The Economist* – must at
some time lead to the emergence of competition due to the
irresistible tendency of *all* operators to extend such a majestic
profit margin over as great a number of barrels as possible –
optimization by volume. The famous $1 a barrel price would
have been possible only if not just the oil company margins
were to have been compressed still further, but if the tax-
gathering OPEC members had trimmed their own receipts
per barrel in the course of taking away business from each
other.

That the then prevailing trend did not continue but was
reversed was due to a number of developments not necessarily
related to each other, but the crucial point of the argument is
that in Adelman's system this was a freakish event, an aberra-
tion, a deviation from an ordained trend, whereas my
approach, based on the likelihood of supply being 'managed'
before competition got out of hand, fully allowed for the
emergence of what American authors still call by way of
oversimplification 'a cartel'. It is thus not correct if Adelman
says in the same letter that he 'failed as badly as anyone to
predict it.'

With hindsight we can see why that time-honoured tendency to optimize by volume has, at a certain juncture almost
three years ago, been substituted by one of optimizing by

margin. There are to my mind a number of reasons for this reversal which I will endeavour to list, not necessarily in order of their comparative importance:

(1) Until the end of the last decade post-war growth and development were not only the prevailing groundswell, but both of these were considered to be beneficial all round; this concept made way for a trend towards introspection and doubts – the Age of Illusion had come to an end.

Environmental fashion led to a questioning of the degree to which natural resources could sustain exponential growth of demand. Even where the excessive imagery of 'Spaceship Earth' – Club of Rome and all – was seen as the humbug it really was, the idea of looking upon fossil fuel reserves with an eye on a longer-than-commercial timespan became inevitably part of oil producing government thinking in general (for very good reasons of her own Venezuela had for some time been orientated in this direction).

(2) A fact had by then finally emerged as a relevant factor: the oil producers had the compelling advantage of functioning simultaneously on two levels – that of an oil operator and supplier and that of a sovereign government. This dual role gave them (once they had appreciated their opportunities) such strategic and tactical advantages as no oil company could match and made it possible to bring about restrictive practices without being subject to any kind of anti-trust proceedings.

(3) Some OPEC countries (especially Kuwait, Saudi Arabia and Libya) were in 1970 and are now in a cash position which makes nonsense of the endeavour to bring forward incremental revenue – there is a point where DCF (Discounted Cash Flow) thinking becomes irrelevant.

(4) The almost fortuitous events in Libya, late in 1970, illuminated the scene like a flash and made evident the feasibility of making the same amount of money (or even more) not by increasing production but by actually and deliberately restricting – or at least containing – output. Once this had been proved – to use one of Adelman's

brilliant sayings, once 'the genie was out of the bottle' –
restriction became cumulatively effective because in a
climate in which postponement of supply is seen to continue to lead to higher prices, such delay becomes progressively more of a rational policy.

For all these reasons the Adelman thesis that the oil shortage may not be real, even if it should prove to be techno-
logically correct, is not as relevant as he thinks it is. He relies
on it being a Natural Law that resources yielding a well-
above-cost return will *always* tend to press forward to the
market at the greatest possible speed (unless precluded by
some machinations which cannot for any length of time withstand the inexorable competitive pressures), whereas my basic
approach makes me define the situation as the *absence of surplus*, even if there is no physical shortage. The Absence of
Surplus admittedly is a conceptual not a technological term –
but it is Adelman himself who has said: 'Belief in fiction is a
fact.'

Adelman is undoubtedly right in his assumption that 'cartels' are brittle, mainly because the more successful they are in
keeping up high margins, the more tempting it is for any
member to indulge in what Adelman calls 'cheating and
chiselling', to try to combine the advantages of high margins
and of an incremental market share. Although there will certainly be no complete absence of competitive pressures, Adelman tends to overlook the fact of the extreme concentration of
reserves within a minute radius centred on the Persian Gulf –
such concentration eases the task of co-ordinated supply programming. In fact one beholds not so much an extreme oligopoly: it verges actually on a state of monopoly with Saudi
Arabia being now the only country where 'instant oil' – the
one which can be made available merely by increasing facilities – is to be had in large quantities. Saudi Arabia on her own
is the arbiter between a state of worldwide surplus and of
actual shortage.

However useful Adelman's critical appraisals are in our
process of economic analysis, there is now the inherent danger
of our 'laying that flattering unction to our soul' in taking at
face value his reassuring statements that all was well or at

least is bound to turn out to be so in due course and that there was now no need to strive (expensively) for increasing independence from OPEC-country oil supplies – in fact only recently he scathingly referred to 'investment decisions made perforce in ignorance' which 'assure massive waste because of the phoney oil crisis.' True enough, the need for the world to spend enormous amounts of by definition scarce capital resources for the partial replacement of low-cost by high-cost energy is nothing short of a tragedy; yet the fact remains that there is no alternative means of reducing to manageable proportions the energy importing countries' subjection to the inevitably monopolistic policies of its crude oil suppliers.

20 International Oil Supply: Notes on Action Now

By early 1973 there were many voices talking of, sometimes predicting in an indeterminate future, an oil crisis. Some voices concluded that the oil-consuming countries, particularly those of the OECD, must present a co-ordinated front to deal with the situation. There was a considerable debate developing as to what co-ordination was required, whether a new organization or a strengthening of existing OECD committees, and for what precise purpose. This is an unpublished note by Paul Frankel, written in September 1973, adding his voice to the debate. A year or so later, and after the price increases at the end of 1973, the International Energy Agency (IEA) was created. There is one element of the proposals here put forward by Frankel which reflects his particular understanding of the company requirements within such arrangements as well as those of government – 'it would be necessary to know that the return for each one of these committed pieces of supply will be reasonably in line with the cost involved and with the netback from deliveries to other countries similarly situated.' This problem, of pricing in an emergency, is still the subject of internal IEA deliberation and discussion with the companies in 1988.

I

It was inevitable that the degree to which the concerted power of the oil exporting countries has become effective in the course of the last few years has made an impact on the thinking of oil importing countries and of oil companies.

Already in the early days of OPEC the question was raised as to whether there should, or could, be a complementary organization on the importing countries' side. The reasons why such a concept was neither realized nor even properly

Unpublished note, September 1973.

formulated were clear cut:

(1) The prevailing competitive climate of the 1960s confined OPEC to a defensive stance and there did not seem to be a need for an endeavour to institutionalize an apparently inexorable trend towards further price erosion.

(2) Since some of the most important consumer countries – the U.S.A., the U.K. and France – had their own particular interests to look after politically and also by way of 'their' international oil companies, they tended to have sufficient stake in the producers' camp to want to avoid identification with importing countries not similarly endowed.

(3) The oil companies thought they could further their own interests more effectively if they were not too closely linked with and overseen by other interests including their own governmental authorities.

When, from 1970 onwards, the growing pressure of producer countries emerged, a search for countervailing power, to be organized by oil importing countries in concert, started but it led nowhere: one country after the other, some of them with almost indecent haste, denied having even considered joint action or re-action to the OPEC countries' pressures; there are some valid reasons for this stance, although the dire threats emanating from oil producer countries of what would happen should the importing countries emulate their own organizational structure would appear to demonstrate that at least OPEC *did* consider coordination of consumer interests to be something which could seriously affect its interests.

However, the objective reasons why an anti-OPEC coalition would now make but little sense appear to be:

(1) The strength of the oil exporting countries' position has been demonstrated clearly in the last few years. Although there is no shortage of potential oil supplies, the balance of supply and demand is so delicate that even comparatively small impediments to the flow of oil to the points of consumption are bound to create a serious problem of adjustment. Since it is now not possible to curtail to any great extent offtake from one source by increasing it from

other sources, importers – even if they coordinated their policies – could not bring effective leverage to bear on any but marginal suppliers, simply because the only way to apply countervailing power, by refusing to take some of the oil available, is not now open to them for a number of reasons.

(2) Although OPEC remains an alliance of very diverse interests and the prospect of encouraging its built-in contradictions remains tempting, it has been realized that the world is really faced not by an OPEC cartel but by Saudi Arabia's virtual monopoly position: only there is the ample potential of 'instant oil', i.e. the one to be had by merely putting in the required facilities; there, and there only, could lie the ultimate decision as to whether there will be, in the next decade at least, ample supply of oil or a real shortage. Now, a monopolist whose commodity one needs cannot be openly confronted, one has to find a way to live with the situation.

II

The stark fact that there is now no scope for confrontation with the highly concentrated power of crude oil suppliers does not mean that there is no need or that there are no opportunities for coordination among the countries which are to a substantial degree dependent on imported energy.

Facing the possibility (it is not a certainty) that producers, primarily Saudi Arabia, will not elect to expand their potential sufficiently rapidly to create a state of surplus, one has to assume that there will be in the coming decade a state of stringency rather than of palpable excess of supply over demand; the main problem is then to avoid a state of affairs in which the several offtakers of oil contest each other's access to limited resources. Such uncoordinated pressures are not only bound to lead to a steeper increase of prices than may be inevitable anyhow, the ensuing friction between the several contestants is bound to embitter their relationships with each other, a process which would be likely to have deleterious effects in the field of general trade, of currency alignments, and eventually of world politics. What is needed is to make sure that the panic which has begun to grip those who depend on oil imports is not allowed to escalate further. This can only be done if all of them can feel assured that it will not be *they* who are left uncovered, simply because others have been quicker or more aggressive. If there could be created a climate in which each country or group of countries could count on not being discriminated against, the overheated temperature would be bound to fall, giving the operators a chance to plan their policies rationally. Perhaps President Roosevelt's famous dictum does now apply: We have nothing to fear but fear itself.

If we thus start from the premise that a search for a modus vivendi, for formulae acceptable to all or at least to most elements in the game is necessary, the question arises what can actually be done to achieve a state of affairs in which at least some of the goals which we can set ourselves could be reached.

It has been generally recognized that a system of allocation would have to be set up in case there should be a fully-fledged supply emergency – OECD is busy adapting the traditional intra-European system to one to include the U.S.A., Japan, Canada and Australia. However vital this line of procedure is, it deals with a *potential* situation, whereas the longer-term and fundamental problem is *actual*: even without a physical shortage, due to a specific interruption of scheduled supplies, there is the underlying problem of oil importing countries drawing from the selfsame pool of supplies, and a limited pool it is, or appears to be, which is virtually the same thing.

There are, to my mind, three levels on which the problems once they have been faced in principle can actually be tackled: they are intercommunicating, in fact they are interdependent, but for ease of presentation they will here be described in turn.

(1) The principle ought to be established on the highest level that the three main offtakers of oil in international trade – Western Europe, Japan and the U.S.A. – will '*respect each other's vital interests*', to use the classical term of international diplomacy.

This involves realization of the simple fact that a concerted policy is possible only if each of the groups (and the

European countries among themselves) realize that, determined pursuit of their own specific interests notwithstanding, there *are* certain points on which the others have claims with which one's own have to be reconciled.

A sound approach in this respect would have to start from an historical position and then one would have to see which amendments are necessary in order to determine an acceptable balance for the future.

In respect of allocation methods there appear to be two extreme starting points:

(a) what can perhaps be described as the U.S. position, according to which the oil which could be subject to concerted considerations, i.e. to some form of allocation, is merely the one coming in by way of seagoing transport, leaving out of the reckoning indigenous supplies and also those moved across land frontiers.

(b) what can perhaps be described as the Japanese thesis, according to which all supplies available to a country have to be brought into the calculation when it comes to assessing the entitlement in each case.

One can assume that a number of countries would tend to subscribe to, or to lean towards, the U.S. approach (most likely Canada and Australia, with the U.K. a potential, on account of her North Sea prospects) whereas Germany, Italy, Spain and most of the oil importing countries in the stage of development would place themselves nearer the Japanese point of view.

Now, it is obvious that neither of the two extreme positions can be tanable: to reach any sort of common ground there would have to be on the one hand a recognition of the fact that countries like Japan, which depend almost fully on imported oil, must have their position allowed for and, on the other, that the need of the U.S.A. to import oil from the Eastern Hemisphere at a rate progressively higher than was hitherto the case, has to be somehow built into the system.

Initially it would suffice to postulate that at the 'summit' of the three groups there should be an understanding, however informal it might be, that one should earnestly endeavour to find a solution to this problem. Only if it is a generally accepted fact that at the top there is a determination to arrive at a mutually acceptable solution can genuine progress be made on the other two levels.

(2) The bulk of the oil moving in international trade is being handled by the 8 'international' oil companies which still happen to enjoy the status of having been the original concessionaires in the Middle East. Any procedures designed to consolidate the situation (and to adapt it to changed circumstances) has to start from that status quo and one has to see how it can be put to the best possible use of all concerned.

Always bearing in mind that in the first instance the task is to reassure those who depend on oil imports, a first reasonable step would be for the oil companies to state to those involved that they would accept the principle that their historical *supply position* in each country would now be considered to be a *supply commitment* with a number of provisos:

(a) The commitment can be honoured only in terms of a share in overall supply available to a company not necessarily in respect of absolute quantities;

(b) Somewhere an increase of supplies to the U.S.A. will have to be allowed for, but on the other hand the non-U.S. customers of American oil companies will have the assurance that the latter are not one fine day going to divert an excessive and unexpected volume of supplies to their home country.

(c) For an operating company to meet the supply commitment it would be necessary to know that the return for each one of these committed pieces of supply will be reasonably in line with the cost involved and with the netback from deliveries to other countries similarly situated. Importing countries, in exchange for a comparatively assured position in the international supply pattern, would have to give a de facto assurance that they will accept the burden of import cost and will not subject their suppliers to unmanageable regulations on grounds of considerations of domestic politics. Needless to say that such

acceptance of import cost burdens by the receiving country involves an obligation on the part of the oil companies
to keep such cost to the offtaker within reasonable
bounds. A great deal of two-way information and of
mutual confidence, to be built up over a period, would be
one of the preconditions for the success of the exercise. If
this can be achieved the governments, rather than be seen
to submit to the unsubstantiated demands of foreign companies, will be able to present to public opinion a mutually advantageous deal: cooperation in adjusting oneself
to a worldwide trend will be justified by a degree of supply
security which no other procedure could have brought
within their reach.

(3) There remains one aspect of great relevance: the developing relationship of operators (and of governments) of oil
importing countries to the national companies (and to
governments) of the main oil exporting countries.

With the share of the traditional oil companies in the
availability of oil progressively narrowing and with that of
the exporting countries themselves widening, the so-
called 'direct deals' will grow in size and importance. The
way these direct relationships are set up and thus will
operate is bound to determine the status of the parties
involved for a long time to come. Thus it is vital for these
new links to be forged in a calm atmosphere of rational
economic assessment of the elements of the deal and not
under the pressure of the fear of being altogether deprived
of supplies required. There are various aspects of the
problem of what could determine the approach of those
directly involved in a way to create such rational climate
of thinking.

The first precondition appears to be a positive settlement of the ways and means of supply by the existing oil
companies: if the procedure outlined under (2) above
worked, the importing countries would know that they
could at least count on getting, via the oil companies,
their share of what was going; in many cases their 'base
load' would be seen to be covered and thus their anxiety
would be brought down to a manageable level. This in
turn would increase their chances of applying some discri-

minating analysis to propositions for 'direct' relationships, instead of their rushing headlong into deals so bizarre and unfavourable as to be likely to mess up their position for a long time to come.

The opportunity for individual operators whose crude oil requirements are largely uncovered (mainly the so-called 'independent' sector including some national companies) to take a somewhat calmer view of their situation can arise only if the commitment of these companies who have still some substantial streams of supply under their control is valid not only in respect of importing *countries* but also covers to some extent those *companies* which in the previous rounds were supplied mainly by the 8 international companies. Nothing has done more towards creating a 'sauve qui peut' atmosphere than the declared intention of the erstwhile major crude oil suppliers to confine as soon as practicable their crude oil supplies to their own refineries. It is evident that such a policy of companies leaving high and dry their erstwhile crude oil customers drives them inevitably on to a path towards catch-as-catch-can tactics.

The repercussions are revealing: not only do prices rise rapidly because all 'orphaned' buyers (and then some others) try to become as early as possible buyers in good standing with the only suppliers with a future; confronted with this buying pressure, the exporting countries cannot help forming the opinion that, contrary to what they had been told previously, they actually will not need the traditional companies any longer; thus they might tend (a) to reduce the comparative advantage which their old concessionaires enjoy by the sum total of their own 'cost' oil and of their buyback prices being well below the market and (b) to accelerate the process of taking away more and more of the available oil from the international companies. In fact, in the absence of an overall understanding, certain importing countries – especially Japan and Germany – could (and probably would) manage to speed up considerably the shrinking of the oil companies' supply base by making it increasingly attractive for the exporting countries to bypass the latter.

Conversely, if – as was suggested under (2) above – the consumer countries could feel secure in their claim on a share of major-company-managed supplies, they would have much less of an incentive to rock the boat in which they themselves were sitting in reasonable safety. They would realize that it is also in their interest to see to it that the transition to a new world-wide structure of the oil industry proceeds at a measured pace and not under panic conditions: it would be fatal if the oil industry as we have known it were to become incapable of carrying out its operational and especially its investment roles before an alternative system of planning and finance has had time to develop. A state of vacuum might lead to a real shortage of oil which the importing countries more than anyone else have every incentive to avoid.

III

Conclusions

(1) The rapidly changing situation in the energy supply structure requires deliberate and concerted planning, simply because individual action designed exclusively to safeguard the particular interests of each individual country or operator will create excessive and self-defeating competitive pressures in the face of a virtual monopoly of supply.[1]

(2) In this process of adjustment the operative components – major internationals, independents and national enterprises – are to play their roles with as little impediment as possible. Such alignment is, however, not possible unless the governments show the way towards cooperation, both internationally as well as in their respective domestic spheres.

(3) A first step in this direction seems to have been taken in the U.S.A. and its repercussion in the rest of the world

---

[1] Contrary to the classical economic concept that higher prices call forth increased supplies, countries like Saudi Arabia may in fact tend to curtail their output if their income per barrel goes up, since this would only advance the point of their monetary saturation.

could become highly significant: within the U.S.A. the existing supply pattern, as far as the several types of operators are concerned, is to be respected, a principle which leads to a certain amount of supply sharing. It can be expected that – as was the case in respect of U.S. import regulations in the 1950s – a makeshift 'voluntary' arrangement will sooner rather than later have to be consolidated in some kind of mandatory system; in such circumstances those American operators who now see themselves uncovered and insecure and thus sally forth to the Middle East and elsewhere accepting any and all terms and prices, will at least know that they will not be too heavily handicapped in comparison with their better-placed competitors. Such a U.S. system would almost inevitably tend to spread to other countries: it could thus become the catalyst of a development which, however rational it was known to be, would otherwise have taken much time and effort to get off the ground.

(4) The approach considered in this paper does not in the first instance involve the formation of any organization. It relies on the component parts of the energy supply systems to follow their own long-term interest and in fact makes it possible for them to do so; these interests, if properly understood, cannot be pursued without due regard to those of others. Whereas the concept, to become effective, does not necessarily rely on hard and fast agreements or regulations, it presupposes sustained mutual contacts between governmental agencies and industrial interests – they are all truly interdependent.

21   The Environment and the Centre

*In November 1973 Paul Frankel was awarded the Cadman Medal by
the Institute of Petroleum. It was an honour that he greatly appreciated
and of which he was justly proud. Its recipients were traditionally from
the larger oil companies and Presidents of the Institute, and he felt that,
with this award, he had achieved not only recognition as an analyst but
acceptance by the oil industry as an oilman. In the Cadman Memorial
Lecture, which he delivered on 14 November, one month after the first
unilateral price declaration by OPEC and in the midst of the Arab oil
production cut-backs and embargoes, Frankel ignored the pressures of the
immediate present in order to concentrate on what he saw as a more
long-term and general concern, the environment. The lecture also reflects
his breadth of intellectual interest and his appreciation for the lessons of
history.*

The previous swing of the pendulum towards intensification of
our industrial life has now come back with a vengeance:
public opinion sees industry no longer as the provider but as
the despoiler. As long as industrial activity was on a modest
scale the environment could take its impact in its stride – there
was plenty of space, air and water around: as long as the
environmental concern remained within certain bounds it
provided an antidote to industrial excesses without affecting
the pursuit of productivity – the industrial momentum took
care of that.

Just as the exponential growth of industry had perhaps
exceeded the limits of its proper place in a viable scheme (or
balanced order) of things, so is the ecological fervour, in its
anxiety to preserve the *environment*, about to pollute the *centre* of
our living (and life), which consists after all of our industrial
existence.

Cadman Memorial Lecture, delivered on 14 November 1973 and published in
Petroleum Review, February 1974.

The ecologists have long since gone beyond the quest for cleaner air and water, they are in fact aiming at a fundamental reversal of the trends which have so far determined the developments of our age. Radical movements usually succeed only if they can make some elements of their programme acceptable to a substantial part of the middle-of-the-road majority of the people – the Communists did it with the tenet of social justice, the Hitlerites by posing as standard bearers of national ambitions. In our case one went a step further: the ecologists' effectiveness is due to the almost unique coalition – an unholy alliance – between two sections of the people which are otherwise poles apart: of the traditionalists who indulge in looking only backwards and of the near-nihilistic movements encompassing a younger generation whose adherents feel that there is nothing to look forward to. Both extremes have united against the middle and are about to engulf it.

The initial reaction of industry in general and particularly of the oil industry was inevitably a defensive one.

Oil was an obvious target for those who criticised the way and the degree to which modern industry impinged on what everybody began to call Environment. Our industry, apart from being indivisibly linked to 'growth', is vulnerable twice over, because production/transportation/refining *and* the use of liquid petroleum hydrocarbons involve a series of emissions into the air, the water and the soil, which are to a certain extent unavoidable.

Because of all this and the kind of chronic bad conscience which oil people developed, they chose to confine themselves to minimising the actual and potential impact of oil on the environment and to playing up the success in containing it, a reaction which reminds one of the man who when asked to return a book replied that firstly he hadn't borrowed it, and secondly he had already returned it. Also industry circles thought that the shadow which the ecological concept was casting upon their activities would go away in due course and that – least said soonest mended – it would be better not to query the antagonists' basic approach nor the conclusions which they drew therefrom.

It has, however, become clearer by now that the oil industry and industry at large is faced not only with a marginal difficulty but with a groundswell of opinion and that we cannot cope with it by evasion or by palliatives but only by analysing the causes of these developments so as to penetrate to the heart of the matter. Bismarck, who knew a thing or two about politics, once said 'One should not try to please the opposition: the opposition is never pleased.'

It is for this reason that I shall endeavour to present what I believe to be at the bottom of what has happened in the last few years during which a protest against our way of life has suddenly welled up. From such analysis we might be able to draw certain conclusions as to the posture which modern industry could, or perhaps should, take up.

I believe that one of these root causes is the inherent fear of change under the spell of which we all live. True enough, familiarity breeds contempt, but to be familiar with the set-up or a stiuation produces a degree of mental security for which there is no substitute. Indeed, one of Freud's neurotic symptoms is the *Wiederholungszwang*, the compulsive need to repeat certain rituals. It may well be that the kind of industrial revolution through which we have passed all through the lifetime of most of us – the revolution which resulted from scientific advance far beyond the dreams of our fathers – has been a bit too fast for our emotional capacities to cope.

There are too many aspects of this reaction to change even to enumerate them in this context, and I propose to confine myself to our reaction to visual change.

The architect's lot is not a happy one: either he remains within the bounds of traditional design and is then considered to be a stick-in-the-mud, unimaginatively wallowing in outdated methods and design, or living dangerously – he might put out something which is altogether new. This will inevitably fail to live up to the expectations of the citizens, will stick out like a sore thumb, and will most likely generate a feeling of outrage. The interesting thing is that when a generation or two have passed, our retina or mental radar has adjusted itself to the new situation and accepts the erstwhile offender as part of the landscape, as it were: when the Vienna Opera House – now considered to be a paragon of excellence – was first built, the pundits and the public were up in arms and one of its architects was driven to suicide.

Now take the skyline: if people in the second half of the last century had been as allergic to innovation as we have now become, the Eiffel Tower of 1889 vintage, which, if you come to think of it, intruded grossly and rudely into the established balance of its post-Napoleonic surroundings such as the *Ecole Militaire* or *Les Invalides*, would never have been allowed to be built; yet lately, when the question of dismantling what has by now become a landmark came up on technical and economic grounds, a shiver went down Parisian spines. (In a similar vein the steam locomotive, that arch-polluter, has become something precious, designed to be 'saved.') Again, Paris as we know it, surrounded and criss-crossed by its famous boulevards, was the result of a ruthless bulldozing of old Paris. If anything similar were to be attempted today Baron Haussmann, the creator of modern Paris, would have been not a hero but a villain. Nearer home – Edinburgh New Town of the 18th Century: for us today the shining example of imaginative town planning would have been condemned as a cold and forbidding agglomeration of uniform mounds of stone impinging on the open space which used to be green and pleasant land adjacent to the original city.

All this adds up to the fact that our fear of change since time immemorial has made us always see the golden age in the past, even if in fact that past was accurately represented by images depicted by Hogarth and Dickens.

One case in point: the changes wrought by the development of our scientific, technological and economic procedures are said to upset the 'balance of nature.' This statement is relevant only if we assume that there is but one such balance, whereas in fact there is (or can be) a *moving equilibrium*: at each stage of development a balance can be established whose element depends on the conditions, the needs and the faculties involved. It would be tempting to plead a case – and by doing so show up all similar endeavours in their true colours – for the maintenance of the dense forests which once covered most of Europe, and to decry arable farming as our first step to perdition – in fact to play backwards the trapper versus settler Western. In the Book of Genesis it is not only said 'Be fruitful, and multiply,' but also: 'and *replenish the earth*' – change is not identical with despoliation.

At rare occasions in history mankind sheds its craving for stability and then glories in a breakthrough towards a new and brighter future – the Renaissance must have been such a period when Ulrich von Hutten cried out *o tempora o litterae, juvat vivere*, which may be rendered as 'This era, these books – what joy to be alive!', and so (through their achievements) may have spoken Benjamin Franklin or Stephenson, Brunel or James Watt. They were creators – they believed they could change things for the better.

Ours is decidedly not such a period. We are suffering, apart from one of the recurrent bouts of self-hate endemic in all sophisticated civilisations, from the hangover typical of the aftermath of great wars. No one could endure the strain and misery they entail if there were not perceived at the end of the tunnel a shining light representing the millennium to follow, and such great expectations inevitably lead to disappointment. After 1918 people felt frustrated by the experience of the futility of Woodrow Wilson's *and* of Lenin's panaceas; after 1945 there was no such disappointment since the war was followed in the developed countries by an unprecedented upsurge of the material benefits, generating the warmth of satisfaction, which the rising productivity of industrial operations had rendered possible. It was only in the late '60s that the hangover set in, when it became obvious that a higher standard of living as such did not assure universal happiness; but what does?

It is there where we have to look for the groundswell which has brought environmental concepts to the fore: the call for clean air and unpolluted water (a consummation devoutly to be wished by any one of us) was not what made the environmentalists tick – it was their inherent desire to opt out of our present-day life – 'Stop the World – I want to get off' was the apt title of a New York musical.

In 1930 Sigmund Freud wrote a short book, *Unbehagen in der Kultur* – Civilisation and its Discontent, from which I quote:

> . . . We come across a point of view which is so amazing that we will pause over it. According to it, our so-called civilisation itself is to blame for a great part of our misery, and we should be much happier if we were to give it up and go back to primitive conditions. I call this amazing, because – however one may define

civilisation – it is undeniable that every means by which we try to guard ourselves against menaces from the several sources of human distress is a part of this same civilisation. How has it come about that so many people have adopted this strange attitude of hostility to it?

And he goes on:

. . . There exists an element of disappointment. In the last generations man has made extraordinary strides in knowledge of the natural sciences and technical application of them, and has established his dominion over nature in a way never before imagined. The details of this forward progress are universally known: it is unnecessary to enumerate them. Mankind is proud of its exploits and has a right to be. But men are beginning to perceive that all this newly-won power over space and time, this conquest of the forces of nature, this fulfilment of age-old longings, has not increased the amount of pleasure they can obtain in life, has not made them feel any happier.

Then he sums up:

. . . It seems to be certain that our present-day civilisation does not inspire in us a feeling of well-being, but it is very difficult to form an opinion whether in earlier times people felt any happier and what part their cultural conditions played in the question.

In the 'hangover' period of ours this underlying self-questioning has made hypochondriacs of us all, and we see ourselves surrounded by dangers and misfortunes which in fact, being self-fulfilling predictions, would overcome us if we were to relinquish our only shield and weapon: our inventiveness and productivity.

So 'we are exhausting the natural resources of our globe': this is seemingly correct, eg in respect of fossil fuels and some other raw materials, but it takes no account of our 'reserves of technology.' On the strength of those we can assume that we shall (a) extend the availability of raw materials by finding resources we do not know of yet, or which we cannot reach on the strength of presently available methods; (b) increase the effectiveness of resources by improving the technologies of manufacturing and utilisation respectively. Even so, are we progressively poisoning our atmosphere, our water and the food we eat; is modern medicine with its reliance on chemical reactions a killer rather than a healer?

There are undoubtedly pernicious incidences but if we are all now being poisoned, how come that we live so much longer than they did in times of old? The fact is that we have developed emotional allergies: the catch-phrase of 'Spaceship Earth' which doesn't seem to be going anywhere relates to a concept of limitations; a concept of opportunities on the other hand involves new frontiers.

DDT is now known to have long-lasting side effects, but has it not virtually eradicated malaria? Thalidomide was a catastrophe, but during the same period tuberculosis, infantile paralysis and syphilis were virtually eradicated. Sulphur in the fuel can be dangerous and lead in gasoline might be: they should be watched and contained, but what if the priority accorded to clean air leads to a shortage of energy? If the feet are cold the nose gets anaesthetised.

The Quality of Life is being impaired by industry's impact: this impression is created by our taking all that industry offers and supplies for granted and *then* looking greedily beyond it: 'All this and heaven too.' We partake of all the advantages of late 20th century amenities, but sated by them would like to transpose ourselves to 'Mansfield Park,' as it were. Incidentally, we need a whole range of National Parks, but can we transform the whole country into one? The world does not owe us a living.

Three items at random: (a) the motor car, the enemy No 1 of all right-thinking spokesmen, also happens to be – with its aura of individually controlled locomotion – the most urgently desired possession of virtually the whole population – *has* it not improved the quality of life? (b) refrigerators, central heating, hole-proof socks and drip-dry shirts, sellotape and polythene – *are* they worth having? (c) the assembly line has become the prize example of soul-destroying industrial practices; we all remember the pathetic figure of Charlie Chaplin in 'Modern Times.' Truly impressive – but would you look to Chaplin to build you an engine? The historic significance of Henry Ford's breakthrough is being overlooked, too: that in the Detroit of the '20s for the first time mass production made it possible to sell the product cheap and yet to pay wages high enough so that the men who built the cars could themselves afford to buy one.

There are people who say that the availability of amenities is badly impaired by the growth of the population and also that the growth of the national product should be inhibited or at least controlled. Now, I believe that people who advocate the zero-growth society, taking it for granted that it will make for tranquillity and for peaceful coexistence of the inhabitants of a country, and for that matter of the world, are basically mistaken. An American friend of mine once told me what the difference was between a dynamic and a static society: in a dynamic society wealth is something to be created, in a static society it is something to be taken away from the next fellow. So we might find that the no-growth society, far from making for social harmony, will do exactly the opposite.

Also we have the problem of the limitation of the number of children, of the elimination of part of the next generation. This goes very deep because the relationship to the next generation is probably one of the few things which, after the weakening of national loyalties and of religious bonds, still manages to take the individual out of the solitary confinement of his own personal sphere, to connect him directly with the world around and with the future before him. I would have thought that a country which would consider children mainly as something to be avoided, rather than something to have and to cherish, would be a sad place indeed to live in. One can perhaps be allowed to indulge in these thoughts if one believes that the population explosion is a horror tale put out by people who are afraid of life; to rationalise their obsessions they elongate the curves and will not see that curves also bend. This one, too, is likely to do so in no uncertain manner – certainly in the industrialised countries – due to the level of education achieved and to more effective techniques of family planning; the cost of living will also do a great deal for a natural revision of those estimates which purport to show that there will only be standing room for people on this earth of ours.

Modern farming methods are being scrutinised. This is as it should be, but if agriculture is not carried on by way of industrial rather than by artisan methods, the world would really be on the brink of starvation. Fertilisers and pesticides (carefully controlled) are essential – compost-grown

vegetables are an elitist fad. *If* we want a chicken in everybody's pot (Henri IV's aim achieved at last) they must be
reared by mass-production methods. I find it a bit difficult to
get enthusiastic about people who are so deeply concerned
about the well-being of the animal which in the next round
they are going to devour with great relish. Such tender
thoughts would hang better on confirmed vegetarians.

We do not like the non-returnable bottles which clutter up
our living-space and reproach the suppliers for them. But
there again, it is not *them* – it is *us*: since there is no labour force
available at meagre wages to carry out the servicing of return-
able containers and since people do want low-priced drinks,
the one-way bottle or tin is the only answer – of course we
must strive for containers which we can dispose of with a
minimum of trouble.

It is interesting to see that the resistance to industrial
procedures – 'people before cars' for instance – comes from
about the same sections of the population which see salvation
in what is now called Community Policies. The call for them
stems from the strongly motivated feeling that our day-to-day
problems are more effectively (and more humanely) handled
on a local level than by the inevitably remote and bureau-
cratic central government machinery. Thus we see a progressive
narrowing of the citizens' focus at a time when the trend of
productive activities goes inexorably towards larger units.
Also if it is the local community which is actually in charge it
will expect to enjoy every possible benefit which can be derived
from the nationwide set-up, yet will inevitably try to shirk
taking its share of onerous responsibilities. An oil-industry
example: our people want to ride in automobiles, but would
you please build refineries in somebody else's backyard. This
is yet another incidence of our attention being riveted on
welfare economics, ie the procedures of dividing the cake
without much thought being given to who will bake it and
how.

It is quite odd that the kind of developed capitalism as it
has emerged in the course of the last few decades has been
dubbed 'consumer society': surely it strove to cater for its
customers, but behind it stood perforce the *productive* capacity.
It would be more appropriate to call a *consumer* society an

order of things which impeded productivity, a society made up solely of (would-be) consumers.

The attitude of the young is most relevant: our youth, biting the hand that feeds them, do not, as I said, consider industry as the provider but as the despoiler; the result is that young people do not go easily or readily or enthusiastically into industry. What can you expect in an atmosphere which speaks of the 'rat-race' instead of thinking in terms of concentrated human effort without which there can be no achievement?

More specifically, when after the First World War I had to choose what I wanted to read at university I became an economist, because I thought that that was where the action was, since it had something to do with the way wealth was being generated. Today, if I were in the same position, I would probably become a sociologist, that is to say, I would busy myself with the disposition of wealth, not with its creation. If we stop creating wealth the escalating expectations of the people will have to remain unfulfilled. Since the bulk of the population is not going to forego easily an improvement or at least the maintenance of their standard of living, any impairment of industrial productivity must have straight inflationary repercussions compared with which our present predicament would look like a picnic.

Actually, semantics are telling as always: the state of 'underdeveloped' countries is due to their not being (sufficiently) industrialised. In industrial countries, however, 'development' and especially 'developer' have become terms of opprobrium. Yet it is the other way round: it is not the shrinking of industry (or its recession into a more primitive or a pastoral stage) which can provide the safeguards against the degree of impairment of the environment, simply because the environment itself is not static: the slogan of a '*post*-industrial society' is misleading: only a *most*-industrial society can generate the degree of sophistication required to combine productivity with an ecologically balanced state of affairs and can provide means to see the task through. The cleaning up of London and Pittsburgh (at a time when ecology was still merely a dictionary term) is a case in point: dirt was cut down and fog eliminated not by reducing heat and power but by using more advanced fuels and fuelling methods.

Environmental concern is a useful corrective of industrial activity – it is no substitute for it.

This leads to the fundamental question whether the very scope of our minds is contracting – seeking the future in the past – or whether we conceive the possibility of, nay the need for, advance in the directions which our scientific and technological faculties are opening for us. I am not alone with this thought: towards the end of his 13-part television pilgrimage Dr. J. Bronowski recently said:

> . . . I am infinitely saddened to find myself suddenly surrounded in the West by a sense of terrible loss of nerve, a retreat from knowledge into – into what?

And he concludes:

> . . . We are all afraid: for our confidence, for the future, for the world. That is the nature of the human imagination. Yet every man, every civilisation, has gone forward because of its engagement with what it has set itself to do. The personal commitment of a man to his skill, the intellectual commitment and the emotional commitment working together as one, has made the Ascent of Man.

The loss of nerve, of which Bronowski speaks, is rampant amongst those who were destined to be imbued by the *spiritus industrialis*. All this reminds us of the title of the famous book of the late '30s by the French writer Julien Benda, *La Trahison des Clercs*, a title best rendered in English as 'The defection of those who ought to know better.'

It is the role of those who understand the mission and the functioning of the industrial process not to surrender its cause to arguments which affect the core of our existence and which go much further than can be justified. There is definite historical evidence that regimes do not collapse under the impact of attack from the outside, they cave in because they have lost confidence in themselves.

22   Topical Problems

This 'topical' was written in the immediate aftermath of the December 1973 oil price increase which, following that of October 1973, had increased crude oil prices by five times in the space of three months. It is a typically Frankel attempt to set this in the context of the rationalizations that were put forward by OPEC at the time.

Any forecast or forward estimate in the energy field has now become virtually impossible because of the very speed of developments: since the 5-year 1971 Teheran Agreement broke down halfway through its intended timespan, it has been said that an OPEC year only lasted a few months. The change between the situation established mid-October in Kuwait and that which has just emerged in Teheran shows that it can now be counted in terms of weeks. Behind this amazing turn and speed of events lies a fundamental change leading to an imbalance of power. Even the pretence of negotiating with the oil companies any alteration of the setup has been shed, the latter have been relegated to the far corner of the field and since mid-October OPEC acts unilaterally and apparently without effective restraint. At that moment the famous words of William Pitt, British Prime Minister, spoken on receiving the news of Napoleon's victory at Austerlitz, became appropriate: 'Roll up that map, it will not be wanted these ten years'.

The extreme price levels of $17 per barrel or so which were reached in an 'auction' for Iranian crude in this atmosphere of shortages, have proved – if proof was needed – that the failure of oil importing countries to agree on a workmanlike allocation of available supplies would inevitably lead to panic buying. The significance of such 'lunatic fringe' transactions lies not in

November/December 1973.

the marginal volumes thus traded, but in the fact that such price levels make those now put forward by the Gulf oil producers look almost reasonable. Although the trend towards higher prices did not originate in the October War atmosphere, the latter, with its strangulating reduction of output in the most relevant Arab countries, must have greatly accelerated it.

There is thus no need to elaborate the fact that the oil producing countries comprising OPEC now have the *power* to impose their will on the operating oil companies and their customers, yet for this very reason it is now more than ever necessary to analyse in an unbiased fashion some of the *concepts* put out by OPEC and its members in support of their claims. Only by doing this can we assess the degree to which OPEC policy statements were justified and ascertain what are their limiations; in the process we should also be able to form an opinion on whether the currently prevailing situation is likely to last beyond the present round.

We shall therefore endeavour to clarify the background and to test the validity of the following OPEC tenets:

I   The price of commodities is determined by the price (cost) of alternative means of satisfying demand.

II   The producer country taxes play only a minor part in the build-up of the prices of products to the consumer.

III Excise taxes levied on petroleum products in consumer countries exceed by far those levied by oil exporting countries.

IV The revenues now derived by the OPEC countries are part of the desirable redistribution of wealth between developed and less developed countries.

I The price of commodities depends on the price (cost) of alternative means of satisfying demand

The long-term prices of a commodity, or for that matter of a service, are positioned between a floor provided by cost and a ceiling which is determined by the price of alternative sources of satisfaction of the demand, and/or by the elasticity of demand, i.e. by the degree of consumer resistance against 'high' prices, a resistance which expresses itself in demand being curtailed at a given price.

The actual price level between floor and ceiling depends on the degree of competition between suppliers – or on the absence of competition. In a state of actual or potential ample supply, sellers have reason to compete with each other for the custom of the buyer, and the more pressing competitive tendencies are, the more closely will the price be drawn towards cost, whereas any reduction of competition will tend to let prices float up in the direction of the ceiling just described.

This picture shows that it is by no means a foregone conclusion that commodities are to be priced according to their replacement value – indeed, such an occurrence would be an exception.

The phenomenon of different cost of the same type of commodity is well known and the supplier of a commodity for which his costs are lower than that of alternative suppliers has the choice of either selling at the competitor's (higher) price and therefore enjoying what is called an 'economic rent', or of supplanting the high-cost competitor by selling at a price which the latter cannot meet. This was, for instance, the situation when, after World War II, oil began to supersede coal in many countries as the main supplier of energy. What then happened was a mixture of the two pricing policies: oil was sold at prices with which coal, especially in Europe, could not compete, yet the difference of cost between oil and coal became so great that the suppliers of the former managed to sell oil which was in the low-cost range (available in massive quantities) at prices which still left them with the very substantial economic rent shared by the oil companies and the producer countries.

If all commodites were sold at the prices of their 'alternatives', the whole wealth of a nation would be concentrated in the hands of the suppliers of *essential* goods and services – e.g. agricultural products – for which alternatives (synthetic foods) would be expensive if they were available at all.

Agricultural supply, however, is not centrally managed but in the present circumstances oil is, simply because of the geographical concentration in a small area of the Middle East of the most prolific oilfields, and by the fact that the

management of the producers' commercial interests is superimposed upon the power of sovereign governments – more
about this aspect further on.

II The producer country taxes play only a minor part in the build-up of prices of products to the ultimate consumer

The OPEC statement is correct as far as it goes. As can be
seen from the figures set out by OPEC and others, the element
of producer country taxes and levies looked comparatively
insignificant at the time these computations were first made in
1964. At that time, OPEC claimed that the producer country
take (then 90 cents per barrel on Arabian Light) was no more
than 6.7% of the average European products selling price,
whereas direct and indirect taxes of the consumer countries
accounted for over 50%. Since then the producer government
take has greatly increased and from the beginning of 1974
appears to amount to about $7 for Arabian Light, but the
whole comparison with consumer prices is not altogether
appropriate.

The real importance of the producer country taxes can be
seen, however, when compared with the *fob cost of crude oil to the
importing country*. What the end user pays depends on the
system of taxation in his country, on the transport and distribution cost and on the local cost of refining. What matters
to the importing country, however, is the *foreign exchange burden
of imported energy* and in that framework the impact of the tax
levied by the producer country is overwhelmingly great.

III Excise taxes levied on petroleum products in consumer countries exceed by far those levied by oil exporting countries

This was true for a long time although it is not true now. To
give an example: on a composite barrel of petroleum products
excise taxes in the U.K. are today the equivalent of $4.2
compared with the $7 per barrel levied by Saudi Arabia on
Arabian Light crude oil from the beginning of 1974. It must
be appreciated that the income derived from the sale of petro-

leum products by the governments of the main oil consuming countries is in fact greater than the excise tax figures show since the net margin of oil companies (which forms part of the price the consumer pays) is subject to income or corporation tax.

Yet the comparison of the two types of 'taxes' is in itself misleading. The taxes in the country of consumption are part of its fiscal system and it is up to the government whether it receives its income by way of excise tax on certain goods or services, or, for instance, by way of a higher tax rate on incomes (direct taxation). The tax imposed by the oil exporting countries, however, is paid not by its own nationals but by those of *another* country.

From a different angle, however, the OPEC argument has some substance: the fact that petroleum products (especially gasoline) can be successfully subjected to high excise taxes is a proof of its price elasticity being moderate, i.e. the traditional price level of gasoline exclusive of tax was considerably lower than that the consumer would have been prepared to pay. In fact the sizeable economic rent which could be derived from the sale of petroleum products (especially those made from low-cost crude) was split four ways: between oil producing countries, oil companies, oil consuming governments – and finally the consumers who used to get the stuff at prices lower than they would have been prepared to pay under duress, as we have since seen all too clearly.

It is, however, far from self-evident that the whole of this considerable 'rent' should belong altogether to the producer country as OPEC has elected to claim; in fact, the size of the rent and its division among the several claimants is determined by supply and demand and by the nature of the competitive situation on either side of the fence.

IV The revenues now derived by the OPEC countries are part of the desirable redistribution of wealth between developed and less developed countries

As in the previously mentioned OPEC policy statements, there is a significant element of truth in this one, simply because virtually all the major oil reserves found in the last decades are located in the sphere of the 'Third World'. Consequently, the transfer of wealth towards oil producing countries does up to a point and in a number of cases provide a redistribution of means which can be considered to be a positive factor on a world scale. The limitations of this phenomenon, however, lie in two directions: firstly the geographical concentration of reserves in a few very small political units (such as the Sheikhdoms) and in some where, irrespective of the size of the country, the absorptive capacity for wealth is limited (e.g. Saudi Arabia) tends to create problems rather than to solve them. The other aspect lies in the fact that a sizable part of the oil exported at monopoly-determined prices goes to less developed oil importing countries, which can ill afford a disbursement of foreign exchange on that scale.

The impact of both these adverse factors could be alleviated if the benefits of the economic rent enjoyed by a small number of oil exporting countries could be spread more widely, in which case the argument at the head of this section would gain in validity. The problem of finding a rational and viable solution lies in the fact that energy, especially oil, has become an international, or rather global, affair which can only with difficulty be accommodated within the confines of comparatively small sovereign-state units. It remains to be seen whether the repeatedly announced steps (designed to alleviate the plight of the developing countries which cannot afford their oil import bills) of which there are so far no actual indications, will now be taken.

Conclusions

Any analysis of the arguments outlined above will point in one direction only: if a watertight cartel or monopoly position can be established by the suppliers of a vital commodity or service for which demand can be curbed only to a limited extent, its price can be pushed up sky-high, especially if the alternatives have substantially higher cost and if, furthermore, their availability is subject to some considerable time lag.

On the other hand, assiduous propaganda designed to show that Middle East oil was too cheap until a few years ago makes the world forget now that even then this crude oil was in the case of the main areas of production sold at 10 to 15 times its attributable cost. This was due to the fact that the companies (and later their profit-sharing host governments) did enjoy their sizable part of the economic rent and the high-cost producers elsewhere (U.S. and Venezuelan oil, European coal) did go along with a policy which spared them the ultimate rigours of competing with low-cost oil.

There does not appear to be an example of a commodity produced on a large scale which would sell at fifteen times its cost, simply because in most cases there obtained varying degrees of competition. The few oil companies which had control of the bulk of the low-cost oil were precluded from abusing this privilege excessively because of the fundamental concept of the need to maintain a state of competition: monopolies (such as public utilities have to be) were acceptable only under strict surveillance by governmental agencies and by the courts, which were to impede abuse of cartel and monopoly power also by, in the case of the latter, limiting dividends to a certain proportion of the amount of equity capital.

The entirely new element in our picture is the fact that the cartel (in which one member, Saudi Arabia, has the decisive monopoly-like position) consists of *sovereign states*. Whereas the between-the-wars League of Nations still stipulated free access to world resources as an essential concept, the United Nations, formed when the empires were about to break up, is altogether based on the principle of absolute and final sovereignty of each and every state. If they are by definition a 'law unto themselves', if there is no supranational authority nor a comity of nations, it is not surprising that, given the opportunity, any state will press to the utmost whatever advantage it happens to have. Furthermore, since the Persian Gulf is in the sensitive area where what can be considered to be the respective overall spheres of influence of the U.S.A. and of the U.S.S.R. meet, there also seems to be no diplomatic restraint which either side, competing with the other for their loyalty, could being to bear on regional elements.

The result is for all to see, a series of price and tax increases has resulted in the tax take alone of the Gulf States amounting now to almost seventy times the cost of production. There is in the energy sector as such no countervailing power – although a workmanlike system of allocation on the international and national levels could have avoided the more blatant excesses of catch-as-catch-can operations designed to secure marginal supplies. In the absence of allocation even a control of prices would have damped the fervour of some of the operators.

Any fundamental re-establishment of a balance of power, however, could result only from the increase in the cost of oil having its repercussion in the economies in general: any widespread recession and/or the patent inability of the international trade and monetary system to cope with the concentration of wealth in a small area of the world, might in the end show to the monopolists themselves that there *are* built-in limitations for them too which they would override only at their own peril.

There is, however, one repercussion of the economic and political pressures brought to bear by the oil states: everybody now, almost inevitably, considers self-sufficiency in energy as a cure-all – this is true enough in some respects, yet what it means in fact is that energy whose physical cost of production is minimal is being substituted by a range of alternatives whose physical cost of production per calorie (BTU) may be a hundred times as high. Whereas the economic rent (difference between cost and price) when earned by low-cost producers is more or less fully re-cycled into the economies, most of the effort in making available energy which has to be produced painfully, i.e. dearly, from less favourable sources is lost to the world for ever. It is this (forced) substitution which will in the end depress the standard of living of the whole world.

23 Topical Problems

A 'topical' that muses on the outcome of the negotiations that finally led, in November 1974, to the creation of the IEA. There is a certain amount of wishful thinking – there was plenty of it around at the time – in the expectation that OPEC would be seeking 'a certain degree of equilibrium'. Frankel also expresses, more strongly than usual, his underlying confidence in the efficacy of governmental direction: 'the main decisions', he says, 'can now be taken only at governmental and inter-governmental levels. The events of the last few months have proved beyond doubt that some sort of international management of energy matters is required'. Well, whatever was 'proved' (a debate of its own) there has still not been any international management of energy matters and many observers would assert that there never will be. Paul Frankel has always had an idealist streak – a necessary ingredient for anyone who wants to create method out of disorder.

The Washington Energy Conference has come and gone and the sequence of events is being recorded later on in this issue. Here we want to trace only the fundamental aspect of the problem which can be boiled down to the simple question: what (if any) is the order of things which could take the place of the oil industry setup to which we had become accustomed?

If we compare the frequency of price changes in respect of other commodities such as cocoa or tin, copper or rubber – and for that matter, of tanker rates – with the overall stability of oil prices in the post-war period, it becomes obvious that the prevalence of a limited number of very large, diversified and integrated enterprises must have been a determining factor.

Stability prevailed because the established operators shaped their crude oil production programmes to a great extent according to their integrated access to the ultimate customer and beyond that to whatever additional outlets could be secured without affecting unduly the market equilibrium.

The eight 'international' companies, so described because of the semi-global scope of their operations, obtained the crude oil in the oil exporting countries on the strength of their concessionary positions, amended as they were from time to time. Due to their interlocking shares in the major concession areas, the terms on which they got the oil were comparable (though not the volume of their supplies) and at least since 1954 (formation of the Iranian Consortium) most negotiations were carried on jointly, or at least in parallel, since it was not in anybody's interest to vary individually the terms involved. Only those not belonging to the original circle (e.g. Mattei's ENI) had an incentive to bid up these terms but only in Libya did newcomers amount to enough to affect the situation.

Thus the acquisition aspect of the companies' operations provided a kind of ceiling which impeded rapid or erratic upward changes in the price level and in the terms of trade. On the other hand, the desire of the concessionaires to maintain a workable relationship with the host countries – and since 1960 the incidence of OPEC – provided a floor below which government take would not fall. In the light of the argument that this mode of acquisition represented a coercive arrangement to the detriment of producer countries two points are relevant:

(1) The tax take of the sixties in the main areas of the Middle East amounted to up to ten times production cost;

(2) In spite of stable tax take per barrel, the actual income from oil exports grew apace with the rising level of demand and with it the transfer of wealth to the OPEC countries increased every year in spite of the receding value of currencies.

On the disposal side, there obtained during the period of the late fifties and of the sixties a mildly competitive market situation which did not exclude occasional sharp bouts of local competition, especially in cases where the newcomers established or tried to establish market shares of some magnitude.

In fact the price reductions in the sixties, due to competitive pressures as far as they went, affected only the companies' margins.

Now the situation is actually reversed: since October 1973 the oil companies are no longer procurement agents of the consumer countries negotiating prices and conditions on their behalf, as it were. These, at least under the shadow of quantitative restrictions due inter alia to the recent Middle East war, are now being almost unilaterally dictated by producer countries. The importing countries, on the other hand, having lost the de facto supply security which the established companies were seen to provide, started to compete with each other for supplies which, they then had reason to believe, had fallen below the level of estimated demand.

The progressive disappearance of the oil companies' power to keep supply and demand in some kind of balance could have different results depending on the supply/demand ratio at any given time:

If it had happened at a time when producer countries vied with each other for a place in the sun, i.e. maximum market share, prices would have gone down a great deal. In the opposite case, at a time of shortage – physical or man-made – the boot is on the other foot and, as we have just pointed out, the competition of buyers made it possible to drive the prices up skyhigh.

The controversy about the justification or otherwise of so-called 'direct deals' has also changed its character: hitherto they were but a deviation from and to some extent an alternative to the bulk of the trade which went through multinational-company channels and which remained the determinant of what was going on. The position is quite different now when direct deals are likely to abound and when it is they which help to determine the going price and also affect the 'buy-back' terms on which the erstwhile concessionaires are being allowed to be supplied.

Hence the rapid changes in the price level, at least as far as the marginal trade is concerned, which as we are now witnessing are not all one way and hence the growing conviction that an open-market situation is no viable substitute for the traditional setup.

The reasons why this is so are of long standing: the risk involved in the initial (exploration) phase and the high capital intensity of all stages of the oil industry always called for security of supply and of disposal. It is the by now complete lack of any means of rational forward planning which is bound to let dry up the flow of investment into all but the most enticing applications. It is the need to re-establish some intelligible and reasonably reliable operational framework which will sooner perhaps rather than later lead to some sort of coordination or rather consolidation of the several interests.

It is in this context that the 'Kissinger Initiative' has to be seen. It has been said not without some justification that in the energy sphere the U.S. interest is not identical with that of Europe or Japan because (a) the U.S.A. is less dependent on imported energy and is potentially selfsufficient; (b) America would not mind her competitors being handicapped by high and yet higher energy costs; (c) the U.S.A. as a superpower has predilections and aims inevitably different from those of lesser operators.

Some of these points are more valid than others, but they are at this moment less relevant than those who put them forward would like to make us believe for reasons of their own. What matters today is that, since nature abhors a vacuum, the world, including the OPEC countries, does stand in need of stability (and of rational behaviour) and that no adequate framework can be established without the U.S.A., in fact such a New Deal is not likely to come about if the U.S.A. does not set the train in motion. It may well be, indeed it is most likely, that Dr. Kissinger has his mind trained in the first instance on his own country's interests, but here (if ever) are they in line with those of the largest part of the world.

The direction to take is clear: appropriate use is to be made of the facilities and the expertise of the operating companies whose continued viability is still required, yet the main decisions can now be taken only at governmental and intergovernmental levels.

The events of the last few months have proved beyond doubt that some sort of international management of energy matters is required: coordination of consumer interests which on their own cannot achieve anything worthwhile can only be a first step logically to be followed by broadly-based contacts with producer countries; confrontation is hardly the right term since no tactics can invalidate the strategic fact that the need of most energy importing countries is greater and is certainly more urgent than is now the need of several of the producer countries for their recompense. Also, in a world obsessed by the spectre of the exhaustion of natural resources, a 'special case' can be made for the price of oil (as there might be one now for the British coalminers).

The tendencies of each of the oil importing countries to 'go it alone' are understandable but a patchwork of bilaterally negotiated and organized agreements, mostly of the barter deal type, do not in their sum total provide a fabric of international viability. They are a meal of several hors d'oeuvres without a main course. On the face of it such agreements appear to solve not only the oil supply problems but also those of paying for the oil by supplying one's own goods and services. There are, however, a number of snags involved:

(1) In the absence of a re-established world code of conduct the terms of these deals are likely to be varied as one goes along by and in favour of the party which happens to be in the stronger bargaining position.

(2) Barter deals involving armaments have probably their built-in equilibrium since, say, Mirages are likely to be sold at 'Posted Price' equivalent. Ordinary industrial equipment and services, however, would be sold at competitive prices, with all industrial nations quoting in order to maintain employment in their respective areas. With oil bought at what is likely to be a monopoly price, the terms of trade for its offtakers would become ruinous twice over.

(3) It is not for us to say here how great are the chances anyhow for the establishment of a viable system of international monetary relationship, but there can be little doubt that the prevailing chaos would get worse still if there should be no rational approach to the flow and the pricing of oil.

All this does not mean that there could or should not be forged a set of more intimate relationships – for instance, that of Europe to North Africa/Middle East. Yet it would be self-defeating if such a move was to be planned in a spirit antagonistic to other links, such as those involving the Atlantic connection. It may be rational for France now to go it alone, thus reaping certain possibly sizable, if not necessarily durable, benefits. She can choose to follow this policy only because she can rely on the other Europeans not to compete in the same way for some favours to be gained and, more important, because she can count on the other eight to save the essence of the Atlantic Alliance and to avoid Europe becoming a prey of potential Soviet ambitions – for the whole of Europe to start treading this path would be first unrewarding and in the end suicidal.

Japan is in a similar position: the fact that she seems to have quietly supported Dr. Kissinger's stand, most likely proves that she has realized where her vital interests lie; for the last couple of years she had tried diligently to establish Japanese enterprises at vantage points, by way of a mosaic of agreements obtained in each case at some considerable cost; but the onset of a deepseated energy crisis has apparently proved to Tokyo that Japan is too big to benefit by pursuing exclusively a backdoor policy. Only a world-wide equilibrium can be her salvation.

And finally who can deny that the developing oil importing countries – of which everybody talks but of whom only few think when it comes to decision-making – will be left in the lurch if the potent industrialized countries try to carve out special deals each for themselves? And who can doubt that a proper solution could possibly be found in an overall plan?

In conclusion we have to turn not only to the likely reaction of OPEC countries to the Kissinger initiative but also to their fundamental interests as we see them.

So pervasive has been the propaganda of producer countries that most oil importing countries have of late thought fit to echo the formers' slogan of the inadmissibility of consumer countries 'ganging up' against OPEC – the French being the most assiduous in this respect, no doubt taking at its face value their own saying: 'Cet animal est très méchant, quand on l'attaque il se défend'.

The fact is that no rational dialogue can be started or

maintained if one side is heavily armoured and organized whereas the other is all over the place.

In fact the oil producers themselves have, when it comes to it, also a vital interest in orderly procedures: the oil countries' unity of purpose, if put to a severe test, may not for ever withstand the stress of political and other rivalries unleashing a new wave of supplies which, by way of old-fashioned competition, may undo the results of their efforts to date.

We believe that it will not be too long before it is generally accepted that a certain degree of equilibrium would be strategically useful, nay vitally important, also for oil exporting countries, even if in the short run the state of general chaos presently prevailing appears to give them some considerable tactical advantage: their wish to maintain this advantage is an exact parallel to the thorough dislike of powerful trade unions for statutory negotiating machinery, such as was provided in the U.S.A. by the Taft–Hartley Act, and more recently in the U.K. by the Industrial Relations Act. The strong are in favour of the freedom to coerce the others.

In the longer run, however, in an erratic situation, not only wide oscillations of prices but also a price level which creates severe dislocation in the whole world and with it not only economic but also political disarray, can hardly be in the interest of delicately poised regimes: their rulers might do well to remember that the legendary King Midas, whose very touch turned anything into gold, had in the end to implore the gods to free him from a privilege which had turned into a curse.

24   Topical Problems: The Value of Crude Oils

The price of crude oil has, naturally to an economist, been a subject of never-ending interest. After the 1973 oil price increases and the subsequent creation of the IEA, the possibility of a producer–consumer dialogue became a matter of intense debate – its desirability, its participants, its purpose, its terms of reference. In the end, and after months of preliminary negotiation, the Conference on International Economic Cooperation (CIEC) took place. It opened in Paris in December 1975 and closed eighteen months later with little to show for its interminable meetings, proposals and counter-proposals. It covered the vast agenda of Energy, Financial Affairs, Development and Raw Materials. This 'topical', written at end-1975, raised some of the problems concerned with oil price which, for many delegations on the OECD side, was the main interest in the Conference.

The onset of the dialogue between industrialised oil-importing countries and developing oil exporters, broadened as it will be by an approach to the problems of commodities at large, calls for an examination of some of the principal facts and concepts involved.

Oil was revalued rapidly and extensively two years ago. Now in the current period of relative calm, the time has come to take stock of the Objective Situation.

I

Temporary deviations notwithstanding, the price of any commodity or service rests on a floor consisting of its long-term supply cost, while its ceiling is formed by the price of alternative means of meeting demand and the limits to the ability or

November/December 1975.

willingness of potential customers to accept a certain price level. Where the price actually positions itself between floor and ceiling depends entirely on the degree of competition prevailing – or on the absence thereof.

The trouble in respect of oil prices in particular results from the difference between floor and ceiling being enormous. To make things more awkward there is also the fact that the costs of some crude oils in massively long supply are very much lower than those of some other crude oils which, for one reason or other, are also to be used.

This conjunction of features results in an extraordinarily wide range of possible prices: extreme competition can push prices down very far towards the floor, whereas concerted action can push them up sky-high.

Because of the potentially self defeating character of outright competiton in a heavy-investment determined industry, there has more often than not prevailed a kind of 'managed' market situation, e.g. the (spelled-out yet ineffective) pre-war 'As Is' Cartel and the (unwritten yet effective) post-war oligopoly, the latter based on the almost complete coverage of the Middle East and of Venezuela by a few companies. However, the power of price determination of both of them was limited by marginal competition and contained by Anti-trust laws and by the pressure of public opinion. OPEC, on the other hand, an organisation of sovereign states, is subject to neither, and therefore can go much further than the companies ever could, whereas competitive forces have so far failed to make much of an impact on the situation.

The ceiling to prices is high: because of the unique and central position of energy in our setup, which reduces the possibility for consumers of doing without; because the world had accommodated only too well to the low-cost oil thus making substitution difficult in view of the massive volume involved; and of course because of the high cost of virtually all alternatives.

There is no one-sided solution: the concept of untrammelled *competition*, where short-term optimisation by way of maximum volume would induce everyone with an ample supply potential to try to steal a march on others and drive all higher-cost alternatives into the ground, is altogether inadequate if applied to a finite resource. It inevitably leads to its excessive use, impeding at the same time investment capable of broadening the supply base.

At the other extreme, the concept of a tight volume-and-price *control* over a vital commodity, which is replaceable only in the longest run and at exorbitant cost in physical effort, proves to be equally counter productive in economic and political terms, if it is used in a single-minded fashion to maximise instant revenue regardless of current and of deferred side effects.

II

The fact that – by concerted action which became feasible only in very special politico-economic circumstances – crude oil prices (in current dollars) have risen almost tenfold has given rise to the idea that in the preceding period they had been kept deliberately low.

There is little to bear this theory out because anyone familiar with the circumstances existing at the time knows that prices were held substantially higher than they might have been had textbook competition prevailed. Thus one cannot say that no Economic Rent accrued to the exporting countries, one can only discuss whether, apart from the oil companies' margin which had become compressed by way of competition, it had been adequate in comparison with the massive Consumer Advantage which undoubtedly accrued to the oil-importing industrial areas of the World.

What is much more doubtful is whether the proposition is tenable that, by not receiving the full value of economic rent now having been proved to be within their reach, the producer countries were really despoiled as it is now widely stated. The fillip which low energy prices provided for the post-war expansion in industrial countries – the degree of its importance has never been seriously assessed – resulted in an ultra-rapid increase in energy demand and the lower-cost oil did accelerate substitution of coal by oil. Apart from the very handsome revenue which the producer countries derived from oil exports even in the Bad Old Days, the overwhelming role which imported oil assumed in Western Europe and Japan provided the base from which the effective pressure could be brought to bear when the moment was ripe for it.

There is a need also to remember that the original tendency of companies as well as of countries to sell as much oil as possible then and there rather than in a distant future was sound economics (and politics) in the post-war period of overriding belief in growth as being A Good Thing, backed as this belief was by oil reserves then considered to be used up in some very relevant instances at a rate which would be sustainable for a hundred or so years.

III

Whereas economic behaviour, determined by the immediate interest of individual enterprises, has the satisfying appearance of intrinsic automaticity, engineered price levels have to be justified. The justification of the price levels prevailing until 2–3 years ago (ten to fifteen times cost in some cases) was probably based on its being a reasonable compromise between the extreme positions described earlier on. The more highly engineered current level results from the desire of some countries to draw to *their* side the bulk of the economic rent/consumer advantage. Geographical and to some extent political concentration has made this possible at least for the time being, and the status of the suppliers as developing compared to that of the main offtakers as industrialised (rich) countries provided a certain degree of moral justification.

Conservation of the resource is another. In a period in which concern about the finite nature of the earth's resources is prevalent, fossil fuel reserves are being looked upon from a different angle than they were in the preceding era of Opportunities Unlimited.

Earlier on in this series of notes on Topical Problems we referred to the concept that finite resources should perhaps not be subjected to the vagaries of market developments and that a system of *rational* management able to take the long view was called for.

It has been pointed out in OPEC circles and elsewhere that the oil-importing countries should in fact appreciate that by the scaling up of oil prices they had been shaken out of their complacency and forced to face the fact that a *replacement* of the currently available low-cost energy would involve considerably higher effort. The earlier this adjustment was seen to be necessary and was in fact made, the better it would be for all concerned.

An offshoot of this thinking is, of course, the concept of a Floor Price – below which energy prices in each country should not be allowed to fall – originally put forward by the U.S. and nowadays assiduously though not always tactfully supported by the United Kingdom.

All this amounts to a number of considerations:

(1) However arbitrary the setting of oil prices may have been in 1973/74, the new level has become part of the landscape since the world has in one way or the other adjusted to it.

(2) Failing a complete breakdown of OPEC – which is unlikely except for straight political conflict among its main governments – prices are not likely to fall substantially in current dollars.

(3) Prices may tend upwards in high-inflationary circumstances, but the fact that the radical upswing of two years ago has anticipated a good deal of inflation will not go unnoticed. Indeed it was Sheikh Yamani who said that the new price level was entirely justified but that it was achieved much too abruptly.

(4) The call for formal indexation of oil prices will be reviewed in the light of the handling of other commodities in the forthcoming deliberations, which are to follow this month's preliminary Paris conference. The idea of formally fixing the development of prices for a host of commodities will almost certainly run into the sands. The experience with the Common Agricultural Policy of the European Community will make for great caution when problems are tackled on a world scale.

(5) The reason why, as we have said now for some time, a prolonged period of continuity of price levels is to be expected, is that *all* concerned have learned to realise that revolutions are really successful only if they lead to a new kind of stability. It may be that still higher oil prices

would not do much further damage; what would be intolerable, however, would be all-round insecurity resulting from arbitrary power being used capriciously.

Perhaps the following passages written on the eve of the 1971 Tehran conference remain relevant:

> Events of the last few years, stemming from the doctrine that a sovereign state can override all commitments it has entered into, have downgraded international intercourse to the level of the jungle. . .
>
> Enterprises – privately or governmentally owned – can successfully operate only within an international framework of rights and obligations from which could emerge a world price level with which both sellers and buyers alike can be expected to live.[1]

---

[1] 'Oil Prices: Causes and Prospects' published as Special Supplement to Platt's Oil Price Service, January 29, 1971 – see pp. 201–2, above.

25 Topical Problems: Oil Industry Integration: Past, Present and Future

Frankel has always been fascinated by the structure of the oil industry, in terms of both economic theory and practice. Ever since Essentials of Petroleum *he has written about it and has sought to test his original propositions. This 'topical' – late 1978 – is another example of his probings into the rationale for integration. It is of particular interest and relevance at a time when the OPEC producer countries are themselves again, in 1988, debating the advantages of forward integration, or at least of downstream investment.*

I

By now the economic motivation is well established for oil industry activities to be concentrated in a small number of large enterprises, and for their operations to cover all, or most, of its phases from exploration right down to end-user marketing.

Size allows of diversification, which in turn reduces the risk of some fatal impact of local (or national) lack of profitability; the spread of such risks, by establishing an average rate of return on capital, ensures a desirable life expectancy of the enterprise. Such considerations are particularly relevant in an industry mainly based on substantial 'front-end' investment, in which the extreme relationship of fixed and variable cost involves risk factors of a high degree.

The tendency of a producer to look for security of disposal by way of captive refiner-customers, and of the refiner to seek the security of outlets by way of direct access to the large number of ultimate consumers, makes for what is called *forward* integration. Thrust in that direction would be less suc-

September/October 1978.

cessful if the marketing enterprise, once it has reached an investment-laden size, would not look for continuity of supply, and if a refiner was not seeking access to assured flows of crude oil to his plant – *backward* integration in fact.

All this is capped by the fact that, since the respective profitabilities of the several phases of the industry are not likely to be uniform, the enterprise which is based on the average thereof is more shock-proof than are those which sink or swim with one only.

Such general considerations were particularly critical once crude oils became available, first in Venezuela and then in the Middle East, whose actual exploration and lifting costs were greatly inferior to those hitherto making up world oil trade (mainly those of the USA), and whose control was almost completely in the hands of seven or eight companies.

The originally very substantial profitability of low-cost oil, to be eased into a market which was price-determined by higher-cost oil and coal, made it highly desirable for the few beneficiaries to achieve an increase or at least the maintenance of market positions – indeed, steps in this direction became wellnigh inevitable.

Thus the postwar period saw the progressive broadening of the downstream positions of companies such as BP, CFP, Gulf and Caltex (the latter especially in the East), whereas companies such as Exxon, Mobil and Shell, well-endowed downstream, strenuously sought to strengthen their upstream positions. On a different scale, the French, Germans, Italians and Japanese, for a blend of economic and power-political reasons, strove, with varying degrees of success, for greater control over the production phase of the industry.

The virtual end in OPEC countries of the traditional system under which the concession-holding company made most of the decisions on offtake and investment has, to a great extent, affected the motivation for integration: if the benefit inherent in the upstream phase now accrues to the oil company only to a limited extent its urge to secure outlets for crude oil which otherwise would stay in the ground is commensurably reduced. And if the erstwhile producer of his 'own' oil becomes a buyer of crude oil, he enjoys a new kind of flexibility, because (subject to some minimum offtake commitments) he can simply refrain from buying if it proves difficult to dispose of certain quantities profitably. Hence even his downstream establishment can afford to become less rigid and more relaxed.

Recently Mr. M. M. Pennell, Deputy Chairman of British Petroleum, said: 'The question remains with us in large companies whether the rationale for vertical integration still exists . . . in consuming areas such as Europe the basis for vertical integration has . . . been severely weakened.'

II

When the OPEC countries took over the management of their resources, becoming not only the de jure but mostly also the de facto owners and disposants of the crude oil, it appeared logical that, having become oil enterprises, they would conform to the classical pattern of oil companies and move along the path of forward integration. In fact, flushed by the initial success and ample cash flow, some of them declared their firm intention of going downstream world-wide; indeed, the Shah announced that NIOC was to become a semi-global integrated enterprise.

Very little progress has been made in this direction and for a variety of reasons:

(1) It was realised early on that starting a network of refineries from scratch, and a commensurate marketing setup, would be unmanageable for competitive reasons in a near-stagnant market.

(2) The acquisition of existing companies proved rather more difficult politically than was originally expected, and the main problem of managerial performance and control took on daunting dimensions once the procedures to follow were more closely envisaged.

(3) The depressed state of refining and marketing, in the first instance in Western Europe, provided little incentive for going there anyway. OPEC countries used to (and perhaps spoiled by) the rich pickings upstream, looked with concern at those sectors being 'in the red'. The traditional OPEC claim for a share in the value added by

the subsequent phases of the industry lost, at least for the time being, whatever justification it might have had: the sum total of integrated profits was lower than that of the upstream phase alone.

(4) The really decisive consideration leading to lack of OPEC countries' initiatives pointing downstream, however, must have stemmed from their confidence in being able to maintain a reasonable balance of supply and demand for crude oil. The main traditional incentive for forward integration was, as was pointed out above, the need to secure and to safeguard a market share for one's crude oil in a situation of ample supply all round. If, however, there should be other reliable means of sharing the market, there would be less incentive for the crude oil producer to seek security of tenure by way of reaching out right to the point of end-use.

A telling example of such a system was the Market Demand Proration according to the Texas Railroad Commission pattern. There, there was the assurance for each and every producer that, within the limits of his Allowables, there would be a refinery to take his crude oil, since overall supply was closely tied to overall demand in the relevant period ('Refinery Nominations'). Just as there was thus no need for the *producer* to establish a Special Relationship to certain refineries (e.g. by way of equity control), any *refiner* could be sure of being supplied according to his requirements and on equitable terms. Consequently, except possibly on the strength of the higher profitability of the producing phase, there was little incentive for backward integration: at the time we described proration as a form of Poor Man's Integration.

Although the several attempts of the OPEC Secretariat to establish what was charitably called Production Programming by its member countries have failed, there appears to be in being a kind of tacit understanding leading to the avoidance of competitive stresses of a magnitude liable to affect fundamentally the general price level of OPEC crude oils. True enough, there are skirmishes, expressing themselves in shifts of differentials between the several qualities and locations, but the centre holds by the comparative flexibility in respect of

volume which Saudi Arabia – and to some extent Iran and
Kuwait – appears able and willing to display and which
provides the basis for comparative stability in respect of price.

All this has been conclusively formulated by Dr. A. H.
Taher, Governor of Petromin:

> The typical downstream investment of the multinational in the
> past was guided by integration to establish a secure market
> position, counting on integrated profits to justify it. As of now,
> such integrated profits are greatly reduced and, in the case of a
> national oil company, may not exist. Furthermore, the security of
> markets for crude oil, in a short market, assumes a different
> perspective. *In other words, for the national oil company, neither of the
> two major considerations in a downstream investment is any longer relevant.*[1]

III

Whereas these considerations appear to have so far inhibited
OPEC countries' forward integration abroad, the establishment of refineries and of petrochemical plants within their
own borders presents a somewhat different set of problems.
These have been discussed at an OPEC Seminar held in
Vienna in October this year. There, especially in the keynote
speeches of Mr. Ali Khalifa Al-Sabah, the Kuwaiti Oil Minister, and of Mr. Ali M. Jaidah, the outgoing Secretary-
General, the pros and cons of industrial policy in this sector
were soberly assessed. The obvious drawbacks involved were
recognised, yet selective progress in that direction was advo-
cated, and the request was voiced for acquiescence and support by oil-importing (industrialised) countries to smooth the
path in this direction.

There always were some advantages in building refineries
near the source of crude oil, e.g. saving of transport cost for
refinery fuel and refinery losses and possibility of serving a
number of different markets out of one refinery which makes
(it) somewhat independent from the particular demand pattern for the several products of any one individual market.

The point that forward integration is economically also

---

[1] In a paper presented to an OPEC Seminar held in Vienna in October 1977. The
emphasis on the last sentence appears in the original.

dependent on the respective profitability of the several industry phases, made above and also confirmed by Dr. Taher, remains relevant also for refineries located in the OPEC countries themselves. The problem is simple: in a competitive market the refiner makes money only if the aggregate of his products netbacks covers the price of the crude oil, actual operating cost and an adequate (and sustained) return on capital invested. Such uplift may be available at any given time, yet its existence is not axiomatic – it cannot be taken for granted.

There is another element to which Professor Adelman once drew attention: OPEC (and its members) may find it still more difficult to contain and control the amorphous products market, highly diversified by qualities and locations, than it is already in the case of crude oils with their floating differentials. One could also restate that, if one fails to sell one's crude oil one can leave it underground free of charge (except for interest foregone) and can lift it later at a more opportune time; on the other hand, refining capacity, not used there and then, is being lost forever. Hence competition between refiners is more acute and sometimes more strident than that between producers – and the price-level changes are sharper and more erratic.

In the absence of clear-cut economic guidance, decision-making will be finally determined by elements of political power. If there should be a consistent crude oil 'absence of surplus', natural or induced, OPEC countries would in fact be in a good position to enforce priority use of their own capacities – e.g. that of refineries and for that matter of tankers, another potential avenue of forward integration. In either case semi-compulsory use of such capacities by the customers would provide for the owners the much needed assurance of their profitability.

For oil-importing countries, however, the sharpened lack of flexibility for finished-products importers may enhance the potential impact of political pressures to be brought to bear upon them: in their own refineries they can use crude oils of different origins; depending, however, on massive products imports, and after they have had to let their own refineries wither away they are twice-over confined to certain suppliers.

IV

If forward integration is by now of lesser relevance for the traditional large oil enterprise, a new form of backward integration has emerged on a significant scale.

In the first flush of newly-acquired independence some OPEC countries intended to go it alone in the field of upstream technology with the help of consultants who were to provide specialised expertise. Realisation of the technological and operational intricacies of offshore exploration and development work – also of those of complicated structures and reservoir conditions – have enhanced the governments' appreciation of the need for such complete and balanced service as can be provided by seasoned operators: a second wind for the traditional oil companies.

At the close of the previous period the companies found themselves with vast upstream establishments which they were loth to dismantle and for which they looked for (useful) employment. In some countries, like Kuwait which had reached its ceiling, there was little prospect of being sufficiently needed to be retained or recalled on attractive terms; elsewhere, especially in Saudi Arabia, the call not for marginal expertise but for essential management was obvious.

In their new role the companies miss much of the erstwhile attractions – substantial profits and reasonable freedom of decision-making – but on the other hand their involvement is also considerably scaled down, especially in respect of direct investment.

The pattern of this current form of maintaining and developing contact with upstream events – a kind of backward integration – consists mainly of initial exploration for which the oil company, in some instances, takes full responsibility, expended amounts being refunded only in the event of a 'commercial' find. Thereafter the burden of investment is to be mainly carried by the host country, the company getting an operational fee for its services and, mostly, the assurance of some offtake rights at preferential or, in some cases, market prices and conditions.

The answers to the question whether such arrangements represent a passing fashion or a valid form of new-style integration, are to be found on more than one level:

(1) The oil companies have to determine whether the initial risk-taking effort can be justified by the prospects of being reimbursed 'cost-plus' in case of success and, more significantly, whether the prospects of preferential supplies justify their risk-bearing involvement.

It may well be that these potential advantages in fact turn out to be something of an illusion: as long as there prevails a state of ample oil supply there is little particular virtue in preferential offtake positions, especially when they show little if any price advantage and involve explicit or implicit minimum offtake commitments; on the other hand, things being what they might be, such prerogatives might be invalidated by an acute shortage of oil. Also, such events would trigger off the IEA Emergency Sharing system and thereby their corporate benefit be diluted to some extent by the general allocation machinery.

There is, however, probably a sizable 'grey area' between the extremes with which the utility of such special relationships would emerge as being important: there might be a prolonged period of supply/demand relationship consisting of an actual, even if temporary, or a potential stringency, well before any persistent shortage was to make itself felt. It is in such period(s) that the benefit of secure supply arrangements would be paramount – especially for enterprises whose highly developed refining, transportation and marketing networks require the assurance of a continuous flow of a substantial volume of input.

(2) The problem arises whether the substantial and (as in the case of CFP in Abu Dhabi) concentrated upstream effort of an oil company can be sustained even taking account of the fact that the company's risk-taking is limited and the financial effort is mainly made by the country itself.

These operations involve some considerable application of the company's establishment and even if reimbursement is adequate, or even generous, these activities are not likely to be capital-forming for the operators – they will thus have to look in other directions for capital formation. Also, they will find themselves left high and dry if and when a particular task has run its course or, possibly, has been terminated prematurely.

The conclusions to be drawn are that this new form of integration is relevant to some extent now, but that it would be rash to consider it as a phenomenon likely to be of lasting structural importance. Integration in areas not subject to an OPEC-type regime, e.g. in the USA or the North Sea, still conforms to a great, though perhaps diminishing, extent to the classical pattern. For the time being the attraction there lies more in the profitability of the upstream phase as such than in the security of disposal and supply respectively.

V

Another significant tendency towards (backward) integration is the perennial quest for an upstream foothold by crude oil-short countries, represented mostly by their state-owned or state-backed companies. This applies not only to France, Germany and Japan, but inter alia also to Austria, Italy, Spain . . .

Their rationale lies mainly in the desire to have direct access to crude oil on whose supply they can rely without interference from others – the wish for a share in upstream profits, although extant, is now in several instances only of subsidiary importance.

One is justifed in asking to what extent this tendency is but a relic from a previous period in which operators had virtual control over the fate of their 'own' oil, and when – especially in the Middle East and Africa – the economic rent attached to this oil was very considerable and therefore attractive in its own right.

Neither of these two advantages still obtains to a significant degree and the sobering considerations which, as set out in the preceding section, apply to the traditional Majors are even more critical also for the national companies of the have-not countries, since the latter do not have the cash flow and the tradition of diversification enjoyed by the former.

There is also the point that many of the prospective targets of exploration are logistically unsuitable for a direct supply function to the initiating country and that such benefits could be reaped only by some sort of swap transactions. In this respect too, such benefit could be reaped only in the 'grey area' mentioned above lying between surplus and shortage.

The conclusion, to our mind, would be that the *corporate* benefit to the respective operating company is highly conjectural and even the *national* one somewhat circumscribed; on the other hand, it is highly desirable that the search for more oil and natural gas everywhere should be encouraged and fostered. Thus, from the *global* point of view, activities of this nature are to be welcomed, but it is fairly evident that such a role is more appropriate for foreign-exchange surplus countries designed to be capital exporters, such as Germany and Japan, than it might be for others.

It appears that the oil-importing countries' trend towards backward integration is nothing but the mirror image of downstream tendencies of OPEC countries; in either case it will be mandatory to create viable structures rather than merely to yearn for status symbols.

Another quite recent phenomenon is the downstream move in the UK and in Norway of state oil companies which have emerged in the wake of the success in finding oil in the North Sea. Statoil has actually taken some steps in the downstream direction which, although so far meeting with a certain degree of frustration, may yet be followed by some breakthrough.

Whereas its Norwegian opposite number appears to favour forward integration in the classical style by way of refining and marketing structures, the British National Oil Corporation, although it would be empowered to do likewise, has so far chosen to concentrate on activities in the bulk markets for crude oil. It is interesting to observe a state oil company establishing itself in the somewhat fluid wholesale market, the more so as much of the oil it handles stems from its Participation deals in which it acquires crude oil at 'market price', a price at which, allowing for some time lag, it inevitably has to sell that very oil.

One has to interpret such a procedure as being based on BNOC's assumption that, in due course, control of certain volumes of oil, once there begins to prevail that famous and much expected state of stringency, will provide, for those who exercise it, a position of power, even if there should be no immediately accountable pecuniary benefit. Contrariwise, in sustained periods of surplus the options held by BNOC would tend to become obsolete.

Looking back at the whole range of forms of integration we realise that by the take-over of certain roles by sovereign governments transnational links have been severed or at least attenuated. Yet the very essence of the industry appears to lead back towards some form of *organised relationship of its several phases to each other* – in fact, integration is and remains its natural habitat. Chassez le naturel, il revient au galop.

26 Topical Problems: Analysis of Confusion

The Iran–Iraq war started in autumn 1980. This had the side-effect of preventing OPEC from holding its twentieth anniversary celebrations (planned for Baghdad in November) and of burying the Long-Term Strategy of OPEC which had been hammered out, under the chairmanship of Sheikh Yamani, during the previous two years. It was by no means certain that the final strategy proposals would have been accepted by all OPEC members but this was never in the event put to the test. The element within the Strategy that was of most keen interest to oil industry observers was the proposal for price and this is what Frankel refers to in the opening sentence of this 'topical'. The piece looks again to the idea that, if not OPEC, then the OECD should be creating a structure for maintaining the stability of oil and energy markets.

I

During the last few months we have been pointing out that the price formula of OPEC's Long-Term Strategy Committee was one-sided, inasmuch as it provided for automatic increases even in times of slack markets but for no ceiling in case of (temporary) shortages.

The underlying intention of the protagonists of the formula was of course to assure importing countries that a flexible supply programme would be designed to avoid substantial shortages, make for reasonable continuity and thus provide a kind of de facto price ceiling.

Yet, even before the OPEC machine had a chance to ratify this concept, the war between Iraq and Iran has blown the whole idea of a regulated system sky high. Whilst we have all along maintained that OPEC had come to stay, as it was too

September/October 1980.

useful an instrument for all concerned to be lightly jettisoned, there would be obvious limits to its cohesion and effectiveness.

Whereas OPEC functioned in spite of the Shah's blatant collaboration with Israel and the deviating attitude of the other members at the time of the Arab use of the 'oil weapon' in 1973/74, a shooting war between two members is something else altogether. The position is rendered more critical still by the apparent split in the Arab lines with Jordan openly and the Gulf States discreetly backing Iraq, whereas Libya and Syria seem to have taken the Iranian line.

At the time of writing it is not possible to see whether there will be a truce, a military stalemate or possibly an extension of fighting, inhibiting further the supply of oil to the world. Even on the most optimistic assumption there is every likelihood that both Iranian and Iraqi oil exports will be negligible in the next few months perhaps even up to one year and in the case of an extension of the hostilities the shortfall may be considerably greater and of longer duration.

Before one can try to assess the medium-term repercussions, it is necessary to realise that the political status quo, on which the oil supply situation had depended, is not likely to be reinstated.

The precarious political equilibrium which prevailed until recently has been shaken and a moment of truth has come.

(1) Whereas the potential impact of the antagonistic superpowers' policies has added to the insecurity in the region, both of them have lost control of what the 'clients' are actually doing. The inherent instability of countries which have become too rich too fast and, for that matter, of those who have remained relatively poor, has come to the fore: with the fall of the Shah a Pandora's box has been opened.

(2) The idea that the origin of instability in the region was only the incidence of Israel, and that all would be well if that problem could be solved, has been shown up as a myth. In fact the antagonism to the State of Israel was the only unifying element of the Arab countries, but even then their inherent disparity has surfaced with a vengeance.

(3) The general belief that the Iranian regime could not withstand military intrusion has proved inadequate. The

regime might eventually founder in the unrest which inevitable hardship will involve, but in the first instance it
has been solidified by the better than expected performance of its armed forces and the addition of patriotic
solidarity to that of religious fervour. And the prospects (if
any) of the return of an Iraqi-related exile 'government'
have been virtually eliminated.

(4) The hold of the regimes in the other Gulf states – tenuous
at the best of times – has become more fragile still.

Taking these elements together the outlook for a reasonably
consistent functioning of the world's greatest concentration of
oil resources is undoubtedly bleak, and the rest of the world
will have to look more intently than has been the case of late
into the need for measures adequate in the circumstances.

II

There is a two-tiered problem: that of the availability of
sufficient supplies and that of the price to be paid for oil
actually or potentially available.

The experience of 1978/79 has proved that violent price
increases can take place in the wake of mere *anticipation* of
shortages, even if there is no actual deficiency in supply. This
lesson having been learned, official propaganda now spreads
the notion that there is no reason (yet) for operators to take
evasive action, since shrinking demand, outsize stocks and the
helpful attitude of some producing countries will prevent the
situation getting out of hand.

The danger now lies in the well-known fact that the first to
believe propaganda (and sometimes the only ones) are those
who have put it out. In spite of the outward calm it is possible,
indeed it is likely, that a period of panic will set in sooner or
later, with countries as well as companies trying to reduce the
danger of remaining uncovered – then it comes natural to
make efforts to get the better of one's competitors by bidding
up the price and/or to make particular efforts, e.g. in the
political sphere, which are likely to be matched by others.
*Self-restraint is possible only if and when each operator (or government)
has reason to believe that its competitors will do likewise.*

The concept that The Market is self regulating and that the price will control supply and demand, however attractive, does not measure up to the impact which politico-military events have on a narrow supply line. Although the current output of tranquilisers may be useful momentarily, the people ultimately responsible must face the real facts and dangers and there is little sign of this taking place right now. There are a number of reasons for this to happen:

(1) The USA, in the last round of a Presidential election, does not want to raise the question of shortages, apart from the fact that its import needs have been reduced and recently there has been nothing from Iran and very little from Iraq.

(2) France, greatly affected by the shortfall from Iraq, and not prepared to swallow its 'go it alone' pride, hopes for the best, and relies on increasing supplies from other sources by diplomatic and other means.

(3) Germany, wedded to orthodox market-economy principles, looks back on a good performance so far – and still refuses to accept the need for anticipatory action.

(4) Britain, now self-sufficient, can afford to keep a low profile and hopes to delay the moment the call may arise to share in the sacrifices.

(5) The Commission of the European Community is acutely aware of the situation but is handicapped by the most powerful members' reluctance to face facts.

(6) The people in the hot seat – the managers of IEA, the International Energy Agency – equally restricted in their mobility by its members, although they are doing useful work in helping to channel supplies where they are most needed, e.g. to Turkey and Portugal, are afraid that, without strong support from the member country governments, any serious endeavours to apportion available quantities in an orderly manner would fail. On the other hand, failure to act timely and effectively, will make the whole existence of the Agency appear irrelevant.

All this tends to add up to what a great writer once called – in another ominous context – A Conspiracy of Silence.

III

We ought now to revert to the actual and potential background of the energy supply position of the world. If we assume that the high post-war growth of energy demand was a temporary phenomenon and that we have entered a slower-moving era, there is no doubt that if all signals were on 'go' for oil, gas, coal and nuclear energy, demand would be amply covered right into the next century even without fundamental technological innovations such as the fusion process.

The trouble is that in spite of widespread profession to that effect, little enough is happening, both inside and outside the oil sphere. The impressive amounts devoted to exploration for oil and gas notwithstanding, such investment, although regionally relevant (and possibly remunerative for the operator), covers but marginal opportunities as seen from the global level: little new is happening in the areas of still greatest promise, i.e. in most of the OPEC countries. Here, as we all know, there are ample reasons for a slow-down, not only of exploration, but also of development work – the oil companies have little if any access and the countries themselves are in a restrictive phase, a tendency which, contrary to classical assumptions, gets stronger the higher the price goes, not weaker.

The medium-term prospects for alternatives are still uncertain. With the exception of France (and possibly the USSR) nuclear developments have run into impenetrable road blocks. There is general agreement on the need for a return engagement with coal, but the massive effort in mining and transportation which would or will be needed is slow in developing. Some of the effort designed to make an oil-like liquid out of coal seems to us (and to many others) a dead-end endeavour, it draws limited capital availability in the wrong directions: coal should be burned (and here great progress in technology is needed and is possible) and if coal progressively takes over electricity generating and industrial heat, the rest of oil uses can undoubtedly be covered easily by naturally liquid and gaseous hydrocarbons: no need to go on a fool's errand by painfully and expensively breaking down the heavy molecules – unless of course there is a political element, as there was in

Hitler's Germany or there is now in South Africa.

It is, incidentally, a fallacy to expect investment in alternative sources of energy to be determined solely by the price of oil. True, the pre-1971 oil price was low enough, and oil availability apparently large enough, to discourage the search for more difficult alternatives, but the present situation proves that there are yet other impediments. The long lead times involved and the magnitude of front-end investment required discourage investors in an era of extreme politico-economic uncertainty. It is not only the unstable international situation which makes us pause before we jump, it is the erratic way our own respective governments determine, and frequently and abruptly re-determine, national guidelines and regulations.

Hence, it is futile to rely on oil prices alone as investment motivation and when risks become too great to be acceptable even to large corporations, governments are the only agents who can face them. They are sufficiently broadly based to take them in their stride and, their policy changes being a major risk factor, the responsibility lies where it belongs.

It is early days to say whether President Carter's Syncrude programme will be a success or failure – the concept of judiciously injecting Government money where it can do most good is undoubtedly a step in the right direction: there are many variants which can be adapted to national and local conditions but the concept of positive co-existence of corporate and governmental policies and management is central in a period when neither of them alone can meet the requirements of the future.

IV

If the medium-term outlook provides grounds for concern, the shorter-term prospects are frightening. The world has lived for some time with the dependence of the bulk of internationally traded oil concentrated on a minute area of potential maximum instability, only things have now taken a turn for the worse and it is more likely than not that further upheavals of one kind or the other will inhibit more of the flow of oil towards importing countries.

For the reasons generally known and briefly outlined above,

there is no effective unity of oil-dependent countries and, if the signs are not read properly and the necessary provisions are not made *before* the event, there can be little doubt that there will be, in the first instance, a further steep increase in price mainly by, what we have called before, the competitive suction effect.

There is general consensus that substantial further price increases will greatly aggravate an already perilous situation. It may be said that another money mountain in the hands of a few small countries would be meaningless since it could not be expressed in goods and services, but its very existence would help to make the present recession look like a picnic.

There is no way to avoid rapid price escalation except by a far-reaching understanding between the oil importing countries *at the outset* of a crisis, without waiting for the famous 'trigger' built into the IEA statute – it would be 'in time to be too late'. To avoid this would involve a sense of acting in political unison as far as the main oil importing countries are concerned. No one who observes their current state of mind can detect the coalescence of policies which would be needed for effective defence of their common interests.

Such unity of purpose was not only the precondition for the regularly called for 'Dialogue with OPEC' which never got going, it is even more necessary in a period where even the mitigating influence of OPEC as an organisation may cease to be part of the game.

Recent events, having thrown OPEC consensus (such as it was) into confusion do not make the West's task easier, in fact they have resulted in a situation in which almost any producer country can singlehandedly upset the precarious supply/demand balance.

The tentative search for a strategy by the OPEC committee had to be put on ice in the circumstances and, although it may be revived at some time or other, it may now be the turn of the OECD countries – always bearing in mind also the interest of Less Developed Countries – to evolve a system which would help to safeguard the world's economic and political stability and by so doing would also preserve and develop the legitimate interests of the oil exporting countries.

27 Urge to Merge Brings Benefits and Pitfalls

As oil demand fell in the wake of the 1979–80 round of oil price increases a new shake-out of oil companies began. A number of large take-over bids were mounted and took place. In this piece, written for PIW, one of the leading weekly journals that has covered since 1961, and still covers, oil industry developments, Frankel looked at this new phenomenon in the context of his general thoughts on integration and oil industry structure.

Mergers and takeovers – agreed or contested – are among the self-adjusting elements of the capitalist system. The management of reasonably prosperous enterprises tends to degenerate into a kind of self-perpetuating oligarchy. Top managers develop and appoint their successors in their own image – hence companies tend to carry for generations the style of their founding fathers. To take two relevant examples in the oil sphere: British Petroleum started with upstream positions, whereas Shell (still called 'Transport and Trading') came from downstream. To this very day each company is still strong in its original habitat.

More often than not such fixation and the in-breeding which it involves lead to some degree of ossification and result in a lack of flexibility needed to adjust to inevitably changing circumstances. One of the results is (or can be) a state of affairs in which some of the assets of the corporation are undervalued because the prevailing setup does not provide a climate in which they can be used to best advantage. Hence it is possible that the actual share price fails to represent adequately the full potential. This, in a different context, may function, or at least look, better and thus makes it possible for an outsider to pay more for the shares than the market will allow in current circumstances.

Early Mergers Recalled

It is worth recollecting that some of the very large oil companies – the Standard Oil Trust especially – grew not so much by grass-root extension as by taking over other companies. These, by being incorporated in the bigger orbit, saw their erstwhile value greatly enhanced. One used to say that they were 'worth more dead than alive.'

These features were particularly prominent in the post-World War II period with the emergence of a number of independent companies with an upstream potential, not sufficiently matched by downstream positions, which were then in the hands of traditional companies much less strong in the 'new' territory of the Middle East. There was first, mainly in 1947, a consolidation among the biggest forces involved, the takeover of parts of Aramco by Esso and Mobil, the Shell–Gulf deal, which lasted a quarter-century until 1974, and the long-term supply deals (at 'cost plus a fee') between Anglo-Iranian (BP), Esso and Mobil. (It is still a tantalizing question why Shell and BP could not make common cause and the former had to link up with a US company, Gulf, to get Kuwait oil which they could have obtained from the British concession partner. It was then assumed that personal antagonism between the two top managements played a role which the then British government either would not or could not overcome.) Some other consolidations followed, such as Texaco which took over Regent in the UK and DEA in Germany, to say nothing of the various affiliations of several companies with Japanese partners.

The unifying feature of virtually all these transactions was the underlying tendency of creating or enhancing integration of the several phases of the industry which was then considered to be the main road to sustained prosperity. This not only assured outlets for one's crude oil or products but also achieved a degree of averaging out the different and shifting profitabilities of upstream, midstream and downstream operations.

Old & New Features

The recent, and especially the current, spate of takeovers and mergers have some features in common with past history. But there are also new elements engendered by the prevailing structural changes of the industry. There was, first, at the onset of this decade, the amalgamation of Du Pont and Conoco and the acquisition of Marathon by US Steel. The former transaction had some justification in its consolidation of oil and chemical activities and might to some extent have been due to special corporate circumstances. The second, throwing together steel and oil interests, whilst not assuring the successful balance achieved by BHP of Australia, might have been due to the attraction of desirable oil reserves in the US and in the North Sea, but was perhaps also because Marathon never managed to evolve a successful image of its own. (When Ohio Oil changed its name to Marathon, observers said that the name was very appropriate since its choice showed that its management realized how far they still had to run.)

What the Conoco and Marathon transactions may have had in common – to some extent shared with some of the current widespread moves – is that in the present economic climate the prospects of middle-sized corporations are none too favorable. They sometimes appeared to combine the drawbacks of the large *and* the small enterprises. Some of them generated into what could be called the 'middle class' in the postwar period of rapid growth, favorable for enterprising management to develop ahead of the pack. In a stagnant or receding environment they failed to develop and to show the staying power which remained the prerogative of the largest and broadest-based corporations. But on the other hand their substantial overheads and commensurate corporate structure had robbed them of the flexibility and nimbleness they had displayed in their original way of life. Yet a number of takeovers represented some logical alignment of upstream and downstream forces, such as the Shell–Belridge, Texaco–Getty and Mobil–Superior deals.

Some of the transactions in the immediate past had a different origin: they were triggered by the intrusion of fringe groups at the margin of the oil industry, developing in the US the concept of royalty trusts. Without an elaborate analysis of such a system, suffice it to say that the endeavor to dissociate some oil reserves from the rest of the enterprise, cannibalize it as it were, though it may provide occasional short-range benefits for some shareholders, does tend to upset whatever balance the enterprise had achieved or which had been aimed at. Nevertheless, an industry trend has developed which has involved some of the most prominent corporations.

Costly Acquisitions

The assault from outside has triggered off the feeling of solidarity – of class consciousness, if you like – of large corporations. It has, for instance, driven Gulf into Chevron's arms, as had the aversion against its original suitor made Conoco to give itself to Du Pont.

Yet there is more intrinsic motivation in these mergers than could be attributed to fashion, which makes all managements rush, lemming-like, in the same direction regardless almost of cost-benefit considerations. It has been said that companies now taking over another company in effect buy reserves at a price lower than would be the cost of finding and developing new reserves of their own.

Only time and experience can show what the real cost of such acquisitions are. The position would be more straightforward if the acquired company was only positioned upstream, thus reverting to the concept of integration. But this is not generally the case – certainly not in the case of Gulf Oil. And it might turn out that the relation of reserves to the rest of the enterprise might not be much different for the new body from what it had been heretofore.

The consolidated company may have to carry on less remunerative refining and marketing activities or, to satisfy anti-trust susceptibilities, it might have to divest itself of assets for which it will receive less than book value, selling in a blatant buyers' market and creating more competition in the process. It may well be that alongside the process of consolidation in fewer and larger units of management, there will be an increasing Balkanization in the refining and marketing phases which, at least for some time, will tend to loosen up the general industry climate.

Big Is Beautiful?

It has been said with a good deal of justification that to put together two corporations with middling or poor management does not necessarily lead to one well managed entity. Yet, there is most likely some merit in the streamlining of the industry which can no longer count on sheer volume increase to cover up its deficiencies, its commissions and omissions. The speed with which these moves were and are carried out owes probably much to the feeling that the current US administration is leaning towards big business more readily than any of its predecessors since Presidents Coolidge and Hoover. There is fear, too, that the climate might change again as time goes on.

It is doubtful that such concentration will affect fundamentally the propensity for investment in exploration once the new companies have found their feet. It all depends on the development of the supply/demand ratio and the expected price level which results therefrom.

The truth which now comes to light is not a new one: operation on a large scale, diversified geographically and in its several phases, is comparatively shockproof. 'Oil is big business – you don't grow wheat in your backyard.'

28   Topical Problems: Is there a 'Just Price' for Oil?

Frankel wrote this 'topical' about a year after the OPEC London Conference at which an OPEC ceiling and member quotas were established to try to hold a marker oil price of $29 per barrel (formally reduced at that meeting for the first time in OPEC's history). The concept of a 'just price' would at that time have seemed desirable to many participants in the oil market – if it had also been the 'right' price. As it turned out Frankel's expectation that price changes would 'be gradual' was not borne out; 'sustainable' they had to be, however painful the process might be.

The concept of a Just Price has been with us since the Middle Ages, if not longer. People have, understandably, tried to find formulae to establish a level of prices 'equitable' for seller and buyer alike. In the pre-industrial climate the guilds of craftsmen considered a reasonable standard of living for themselves as their due, an idea which seemed workable in the somewhat static setup of the town and of a surrounding countryside ruled in a feudal fashion.

The higher mobility of the oncoming industrial age substituted the traditional concept for that of prices evolving from the Market Place – resulting in ruin and misery for some, compensated by gains and prosperity for others. The signals emanating from the supply/demand ratio were considered to be the proper guidelines for economic behaviour, rooted in investment decisions.

At the onset of our own century – and obviously first in the USA – public concern mounted that the very size of some enterprises – and worse still the combination of such entities – could help in manipulating market events in favour of

January/February 1984; also published in *Platt's Oilgram News*, 30 March 1984.

suppliers and to the detriment of consumers and thus of national interest. It is no coincidence that the term Anti-Trust was fashioned to the measurements of the oil industry and its first large-scale concentration by way of the Standard Oil Trust.

On the other hand the collapse of the oil price level in the early thirties, when the deep economic depression happened to coincide with the discovery of unprecedented large reserves in East Texas, led to governmental action to contain supply so as to buttress and maintain a price level which was considered to be adequate – or just.

I

The problem is still with us – with a vengeance.

Since the advent of oil in the Middle East the problem to which we keep on coming back, i.e. the wide range of production costs, has made it more critical and yet more intractable.

If there is, or should be, a free-flowing world market – laissez aller, laissez passer – the low-cost oil would supplant that which is higher-cost: there is, or could be just now, enough of the former to make recourse to the latter unnecessary. Don't let us forget that, as per David Ricardo, the high-cost variety of any commodity determines the price only if 'it is needed to meet demand'.

The trouble in our case is simply that the full reliance on lowest-cost oil is unacceptable for a number of reasons:

(1) If only the lowest-cost oils were to be drawn upon the concomitant increase in production could only be sustained for a limited number of years in many instances. And well before then the impending scarcity would tend to drive up the price level.

(2) The exclusive reliance on a small number of points of origin would mean a critical political dependence on possibly unstable or unfriendly areas and could involve also problems of the international flow of currencies.

(3) The virtual exclusion of higher-cost resources would create grave problems of unemployment of both people and of existing and future investment.

All these self-evident repercussions prove clearly that a 'free' market, orientated solely towards the lowest available cost, cannot be a workable alternative.

As an alternative it could be envisaged to establish a series of protective systems, akin to the import regime extant in the USA for several decades.

Such regimes would:

(a) transfer to a great extent the 'rent' now enjoyed by the low-cost producers to the authorities of the oil-importing countries and
(b) by eliminating, or at least reducing, the price advantage of low-cost oil would safeguard the tenure of alternative (indigenous) higher-cost alternatives. Dr. Kissinger's Support Price idea was a stab in this direction.

To set up such a sophisticated and involved system – not too different from the Common Agricultural Policy of the European Community with all its rigidities, flaws and drawbacks – there would have to be a high degree of consensus between virtually all oil importing countries and in each one of them a desire to determine affairs altogether by governmental decree, a desire which is, at least in this round of the game, patently absent in most of the prominent countries involved.

Thus, with no obvious response to the problems with which the world is being faced, we are bound to analyse the impact – positive or negative as the case may be – of the several alternative directions in which developments may take us.

II

Perhaps it is appropriate to review first the impact of a collapse of oil prices in the wake of the inability of the OPEC members to maintain the necessary degree of solidarity. It is obvious that it is in the interest of producers in general, unless they have limitless reserves in relation to any conceivable level of potential demand in the foreseeable future – and even more so in the case of many of the OPEC members – to 'optimise by margin', i.e. to contain total supply in the face of stagnant demand. Yet, contrariwise, there can emerge at certain moments an overwhelming urge for countries with a large population and immediate budgetary needs, to break out of the restrictive system and, for however short a period – 'optimizing by volume' – steal a march on their competitors.

Whereas a total collapse of prices, say to $15 per barrel or less, would place severe burdens on the economies of many producer countries, there still remains the question whether, and to what extent, other countries would derive some considerable benefit. It has been pointed out that, if the two price explosions of the '70s have been instrumental in bringing on the deep depression, a lifting of the incubus of the high cost of imported energy would help the world towards further and more rapid recovery.

There is no doubt that the foreign exchange drain would be greatly reduced for countries which are not or only moderately self-sufficient – in the first instance that would apply to a number of non-OPEC LDCs but also to countries such as Japan and Italy and to some extent to France and others. Britain and Norway, however, and also Mexico would suffer immediately and directly.

The point, however, on which more research would have to be carried out before a valid judgement can be arrived at, is whether *any* rapid and far-reaching change of the price of a vital commodity – oil being still the biggest item in international trade – inevitably has disrupting features. There is also the experience to take into account that violent moves, up or down, carry within themselves in due course a like move in the opposite direction.

The world gets used to almost any status quo and it is the need to adjust to a fundamentally changed setup which creates most of the difficulties which we have encountered.

III

Much has been said about the merits of price stability – a concept about which we have always had our reservations. Nothing can be maintained stable in a continuously shifting and changing world. What it is possible to aim for is continuity. Such a state of affairs prevails if and as long as changes take place gradually, with any one of them, within a reasonable time span, not being so great or deep as to disrupt altogether the existing setup.

Markets, left to their own devices, act and react under short-term auspices and, as we have pointed out before, the comparative continuity of oil prices was due to their having been managed by someone or other for well-nigh a century.

The recent violent events, however, have taught us that the industry has after all some degree of self-adjustment, a faculty which was not evident as long as the changes took place within a moderate range.

Under the impact of explosive changes, the price elasticity of demand, previously negligible, became operational and alternatives to over-priced and politically insecure oil – coal, natural gas and nuclear power – did appear on the scene in substantial volumes and conservation became a relevant factor.

Although all these reactions were slow in emerging we can now see that in the long run there is a ceiling beyond which oil prices cannot rise for any length of time without curtailing its output severely and to some extent possibly permanently.

Thus, if there are limits to the 'management' of the industry, if there are floors and ceilings, as it were, this would be proof that it is possible to combine the respective influences of straight market factors with a more deliberate design of market *structure*.

Such a dualistic system did in fact prevail during most of the period when some degree of market control was exercised by a few large enterprises: subject to legal impediments and to public opinion such companies could not easily dissociate for any length of time their behaviour from the underlying supply/demand relations which also influenced their individual competitive stance. It was only when politico-military events gave sovereign governments a chance to 'take off' altogether, that prices went up sky high. Now the limitations even of governmental power have been shown up and the dogma that setting prices was the sovereign prerogative of producer countries, which we have never accepted, has been proved to be an empty boast. When the concept of Permanent Right To A Country's Natural Resources was formulated, we pointed out that if the terms at which these resources were made available got out of step with what the market would bear, their control would in fact become 'permanent', the oil remaining in the ground for ever.

At a time, only a few years ago, when it was endeavoured to establish a system under which the rates of exchange of the several currencies were to be de-controlled, but changes in any given period were to be kept within certain limits, there was being used the term 'Crawling Peg', indicating only slow-motion progress in the direction which market forces dictated. Not much came of this exercise, but the concept does have general validity, applying most likely also to the current problems of the oil industry.

It can thus be surmised that some downward revision of a price level, which had reached dizzy heights at a period of panic, was not unjustified and this is what has in fact happened in the course of the last one or two years. The decrease, however, due to a certain degree of consensus among OPEC members and to a 'helpful' attitude of the main non-OPEC producers, was a gradual and a limited one. It is likely, but by no means certain, that this process has not come to an end, provided there are no further outside events which would interrupt or diminish the flow of oil from the Gulf area.

IV

In a world of great complexity, with many contradictory interests, every one of them involving claims, legitimate in themselves, there cannot be any one 'just' price for oil.

Although the rent enjoyed by the low-cost producers is still enormous by any standard, by prevailing now for a decade high prices have become part of the economic and political landscape and a complete and sudden reversal might create almost more problems than it would appear to solve. On the other hand careful and gradual adjustment, which pre-supposes continued coherence in the OPEC camp, sympathy of non-OPEC producers and some degree of acquiescence worldwide, might in the circumstances lead to a 'workable' price level.

The time for a formal dialogue between the main interested parties has not (or not yet) come. The experience of the elaborate conference in Paris only a few years ago has proved the futility of word-marathons, where each party is bound to spell out their respective maximal programme, and to lose the flexibility which only quiet diplomacy and matter-of-fact action can provide.

What would have to be established soon is a high degree of mutual understanding and a certain parallelism of action and reaction. Much more work is still to be done in an uncommitted way, to identify the economic and political repercussions of any one of the several paths which would appear to present themselves.

As we have shown it is possible that the current price level will be abandoned in favour of a much lower and in certain circumstances of a sizeably higher one; the odds are, however, that changes, as they arise, will be gradual and thus sustainable.

29 Time to Redefine Market Forces

Paul Frankel gathered together what became known as the Frankel Group in 1984. Participants were Robert Belgrave, Umberto Colombo, Etienne Davignon, André Giraud, Toyoaki Ikuta and Ulf Lantzke. Frankel was motivated by a compelling need to try to force policy-makers to react in what he felt to be the only rational manner to the troubles which already beset the oil market and which he was convinced would become more acute if nothing were done. The group issued three statements – in 1984, 1985 and 1986. That of 1985 was the most explicit and strongly argued. Its most important message and plea was for governments to get together to solve problems through mutual discussion; they looked for co-operation based on interdependence rather than the atmosphere of mutual belligerence that they perceived. It was a 1980s version of Frankel's 1940s analysis in the Essentials. *The phraseology was updated but the thinking was the same. The appeal was not heeded. It was not, it seemed, what the policy-makers, nor even the market-place, wanted to hear.*

It has become a cliché that the trade in crude oil and products is now dominated by the spot market and that spot prices are persistently weak and tend to pull down official government selling prices of crude, whether those governments be in Opec or in Europe. It is widely accepted that this tendency is to be applauded and brought about the 1982–1984 reversal of oil supply/demand relationships – and also that repercussions in the pricing realm are a complete success for the 'free market economy.' Yet, like most clichés, this one obscures rather than illuminates the reality, and closer examination reveals a different and more complicated picture. That prices depend on supply and demand is obvious. The way the latter two are

The Frankel Group, March 1985 (published by *Petroleum Intelligence Weekly*, April 1, 1985).

formed, and the way they interreact with prices, is not. Nor is it obvious who or what determines the determinants.

The first thing to be said is that the much heralded fall in oil prices may turn out to have been a mirage, brought about by excessive strength of the dollar. As the new head of the IEA pointed out, prices in European currencies have risen – by 13% – since 1982. A similar situation appears also in Japan. That is what matters to European and Japanese oil refiners and consumers, and it matters also to crude oil exporters, who see demand for their oil reduced in consequence and whose purchases are by no means confined to dollar sources. Within the US, it is true that prices have fallen. Yet even here there appears to have developed two views as to whether these lower prices are likely to last long in a split market. Investors would not continue the search for high-cost oil, as they are doing, both within the US and elsewhere, if they did not believe in a long-term market that would remunerate their investment.

Reinterpreting 'Market Forces'

What is needed now is a reinterpretation of the term 'market forces.' Traditionally, the bulk of oil used to flow through integrated channels of major oil companies and by way of long-term contracts. On the fringe, there were some short-term movements, mainly in what was called the 'independent' sector. One used to say that these transactions related to the 'collapsible extension to the main structure.' With the ending of upstream control in Opec countries by the oil companies, and the more recent onset of a buyers' market, the advantage of 'owned' oil and of term contracts was eroded and, with eager supply exceeding shrinking demand, the wheel turned full circle: the 'spot' price became the relevant item all across the board.

Like all commodity 'spot' prices, the market depends on the immediate supply/demand relation, which may fluctuate widely and frequently. This type of market has an intrinsic relevance for the operator, yet it provides little guidance to the investor, whether in a finance house, the World Bank, or an oil company. Oil companies in particular still appear to put a premium on owning some part of their crude requirement, especially if this can be within the OECD area where some degree of 'sanctity of contract' still obtains. They are able to do so in part because of an implicit assumption that respective home governments are prepared to provide them in one way or another with a protected market in the interests of national supply security. Yet from a world supply viewpoint, existence of de facto protected markets for oil produced within the OECD will not for long ensure continuity, because the OECD as a whole cannot do without access to Middle East reserves. But in present circumstances neither foreign investors nor Middle East governments are prepared to put money into production and transport facilities that will be needed in 10 years time. If this situation does not change, then the present glut will inevitably be turned into a shortage.

New-Model Contracts Needed

There thus emerges a need for two different types of supply contracts, one which reflects the fast moving, short-term market and another which takes a comparatively durable supply/demand ratio into account. Such differentiation has existed all along in tanker transportation. There was a substantial difference in approach between time charter and spot transactions, the former being fixed for a sizable period of time (although mostly subject to escalation related to general changes such as currency fluctuations and cost), and the latter which reflected – and created – short-term freight market developments. Another relevant aspect of freight contracts was the AFRA concept, which provided for periodical assessments of all charters extant on a certain date. These were consequently affected by short-term developments, but reacted more slowly, due to the impact of more solid long-term elements.

There is no such thing as market stability in an inexorably changing world, and it is in fact 'continuity' which is called for – a system in which changes take place slowly, without being of a magnitude which would shatter the existing setup altogether. A real price collapse would do a lot of damage to the world energy system, and a price surge would inevitably follow which in turn would again heavily damage the world economy. To avoid these, there is need for reestablishment of a long-term relationship between producers and offtakers. In order to bring this about it would firstly be necessary to convince market participants, including consumers and governments, that everybody stands to lose if the world supply system does not regain some degree of continuity. Secondly, confidence will have to be renewed among market participants. Difficult as this may be, given the experiences the world has gone through over the last two decades, it should be obvious for everybody that nobody gains from a situation in which oil is not regarded by many consumers as a reliable fuel any more. Thirdly, it should be possible to devise a new type of contract between main suppliers and offtakers which would cover a substantial oil flow. To provide flexibility within an agreed range of volume and price, such contracts would involve a floor and a ceiling both for price and volume (preferably not too far apart), within which there would be escalation or deescalation according to shorter-term criteria. Such commitments, if entered into in good faith, combined with an improved understanding of the consequences of breach of contract, could provide a substantial security element for suppliers as well as offtakers. It would introduce a degree of market leadership by those far-sighted enough to enter into them, which could not fail to have a sizable impact on the rest.

A prospect of continuity would also facilitate investment planning in exploration and production of crude. However, the refining sector may well remain more volatile and less suitable for longer-term commitments given the degree of overcapacity prevailing in this sector. Yet, even here there is discernible groping for less volatile price levels.

Umbrella Agreements

It has to be recognized that the type of purely commercial contract envisaged above may be overturned at any time by the sovereign power of the exporting countries, whose governments have in the past used unilateral application of the 'changed circumstances' doctrine, either to raise prices when they believed they were in a seller's market or to pursue some political objective. The position is further confused where governments have acquired commercial company interests in addition to their regulatory and political role. Even in the US, regulation of trade in the interests of public safety, competition and the environment, as well as national security requirements, represents a high degree of governmental 'involvement' in the energy business. In the UK, the very existence of BNOC involved the government in pricing, as does Statoil in Norway.

Most contracting parties in the crude oil business live in different countries, and political conditions vary. The problem is aggravated because in many cases a contracting party is itself a creature of its government. Expectation of continuity under these circumstances is difficult to imagine, unless the governments concerned can provide this. One way could be by agreeing amongst themselves to provide a climate in which the legal doctrine of 'pacta sunt servanda' (contracts are meant to be observed) is the rule, or – as a minimum – to provide by means of an 'umbrella agreement' a deterrent against breaches of contract. Of course, such understandings imply a limitation of sovereignty, as do all international agreements. It may well be that this might prove to be a hopeless task in the case of oil. But it is a characteristic of leadership to keep on the lookout for a favorable conjuncture which can make the desirable become the feasible. The present situation – in which Opec countries are feeling the pinch of reduced demand, a threat to prices, and a danger that oil might develop into a secondary role in the overall fuel mix, while OECD countries see a need for continuity in order to sustain their own internal investment in high-cost energy as well as provide overall economic stability – might provide such an opportunity.

Possible Dialogue Partners

To get things started, there is need to identify a valid interlocutor on either side. Neither Opec nor the IEA is qualified to do much more than exchange information on supply and demand and their future prospects, though this might be valuable in itself. With the US strongly stressing pure market forces in this particular field on the one side, and Opec split by political divisions on the other, the bodies with both capacity and incentive to initiate serious discussions are the European Community or Japan and the Gulf Cooperation Council. If these bodies could start a serious evaluation of the feasibility of some sort of non-exclusive umbrella agreement, to refrain from overturning commercial contracts commercially arrived at, that might give producing and importing companies, whether state or private, the necessary confidence to enter once again into long-term contracts, and might open the way to wider understandings. In short, the main obstacle now to a stable system is lack of confidence.

The time is probably ripe for some initiative to introduce confidence-building measures into the relationship between crude oil suppliers and oil product consumers. Within a single country, with its own crude supplies, refiners are the only necessary intermediary. But, internationally, governments or coherent groupings of governments should now play their part.

30 Postscript

For well over thirty years these notes have been included in
Petroleum Economics' regular feature 'Oil Industry Developments', sent out to its clients, originally bi-monthly, more
recently quarterly. For many years they have also been mailed
to a sizeable number of friends all over the world. I have
written all of them, never having missed an issue.

Now, this will be the last time these notes are being issued.
Some considerable time ago I decided to withdraw from the
oil scene, in which I have dwelt for more than three scores of
years, when I shall have reached the age of 85 years, ending
up if possible with a bang, rather than a whimper. Looking
back at the whole series, I can see that it is like Hamlet's
Players 'the abstract and brief chronicles of the time' yet the
running commentary is imbued with the basic industrial philosophy of mine, as it was first formulated in my 'Essentials of
Petroleum', published more than 40 years ago. Thus I may be
allowed now to restate some of these basic tenets and to give
some thought to their validity or otherwise now and hereafter.

I

In my formative years the oil industry climate was determined
by the control of supply by a small number of very large
internationally integrated corporations on the one side, and
by governmental impact on the other. In today's political and
industrial climate, with its tendency towards unfettered competition and its revulsion against governmental interference,
both tendencies look altogether inappropriate, yet I still consider that they both form an essential part of what makes our
industry. To take the role of governments first: those who
vociferously oppose it in industry and commerce, insisting
that 'you cannot buck the Market' forget that this Market, as

we know it, is now determined altogether by the Anti-Trust philosophy and practice – failing this governmental veto the market's shape and procedures would be quite different, tending to be dominated by combinations of a higher order, whereas there is now strong pressure for a more atomistic structure.

Furthermore to organise a community's life altogether according to the signals provided by each operator maximising his revenue in purely monetary terms would lead to entirely unacceptable social and thus political results. Thus it is inadmissible to talk of governmental 'interference', a term that would only be applicable if the economic system *on its own* could establish a viable existence of the community. Since this is obviously not the case, some co-existence – peaceful or otherwise – of the two systems remains inevitable. Whereas this 'cohabitation' is called for at large, it has its particular significance in the energy and oil domain: not only because of the all-pervading relevance of energy for the community, unparalleled by any other commodity with the exception of food, but because of the excentric location of source and consumption respectively.

We have long since forgotten that the first oil wells were completed in the North-East of the USA, in the heartland of its industrial life. The move to the South-West, Texas and Oklahoma, then belonging to the LDC States, changed the demographic and political balance of the USA, yet for almost half a century the United States as a whole was not only the biggest producer but also the main consumer of oil.

Since the time of the first World War, the new oil resources emerged first in Venezuela and subsequently in a massive way in the Persian Gulf. Ever since oil became inevitably a matter of international concern. No government of exporting and importing countries alike could thereafter fail to involve itself, directly or at armslength, in oil affairs.

As early as July 1913 Winston Churchill, then First Lord of the Admiralty, concerned about the dependence on fuel oil for the updated Royal Navy, said in support of the acquisition of the Anglo-Persian Oil Company (now BP):

> Our ultimate policy is that we must become the owners, or at any rate the controllers at the source of at least a proportion of the supply of natural oil we require.

It is in this sense that after the last war I modified Clemenceau's famous dictum about war and the generals by asserting that oil was too serious a matter to be left to the oil men.

II

The other crucial signpost, to which I have referred time and again, is our industry's RELATION OF FIXED AND VARIABLE COSTS.

The fixed costs, relating to investment in all phases of the industry – exploration/production, transportation, refining and marketing – are of a very high order, whereas the variable cost, depending on the degree to which installed capacity is being used, is comparatively insignificant.

This extreme relationship makes the industry basically inflexible: the classical assumption that higher prices call forth rapidly more supply and reduce demand apace, whilst lower prices lead to rise in demand and simultaneously to contraction of supply, all this automatically tending to re-establish a balance of supply and demand, fails to apply to this industry.

Price elasticity is in a low key all round: price rises are slow in calling forth new investment, partly due to the inevitable time lag. Falling prices in fact tend to increase supplies for a while, since the last barrel carries only nominal cost and there is a need to maintain overall revenue. Demand too is somewhat inelastic, especially for motor fuel, since the investment in the vehicle cum taxes and insurance makes it illogical to reduce mileage because gasoline has become a few pence dearer. It must be admitted though that nowadays the price elasticity of fuel oils has risen considerably by the latent or actual competition from coal, natural gas, nuclear power and by conservation. What still matters though is the fact that the oil industry – as distinct from the classical model mentioned above – is not self adjusting and thus since time immemorial has called forth forces willing and sometimes able to manage supply in a way designed to bring it in a workable relationship to demand.

III

Another of my hardy perennials has been the concept of large managerial units developing because their ability to average out different profitabilities by way of geographical diversification enhances their life expectancy.

Another aspect of this phenomenon is the tendency of straddling the industry's several phases by way of vertical integration. Since the respective profitabilities tend to be different, to be in more than one or in all of them, again makes the management particularly shockproof. The main feature, which has all along pushed oil companies towards integration, because it provided security of outlet (Forward Integration) and, on the other side of the picture, security of supply (Backward Integration) is still of crucial importance. The main oil exporting countries have now cottoned on to this concept – having taken on the role previously played exclusively by oil companies, they cannot altogether avoid following the paths trodden by the oil companies ever since the early days of the industry.

IV

In the '30s and '40s – the industry was fashioned by two ruling forces: the concept of the few outsize internationally operating companies – the Seven Sisters – and in the USA, where the Anti-Trust policies excluded any non-governmentally backed cooperation, by the Roosevelt-inspired Market Demand Proration. The latter was a cartelmaker's dream – each month the refiners would table their demand, which was met by a system of Allowables, embracing all producing wells which automatically precluded either surplus or shortage. In the rest of the world the system of interlocking concessionary positions assured a reasonable balance of supply and demand. The end of prorationing, due to the USA surplus capacity having disappeared and the demise of the traditional concessionary system, left a void, with no one being in a position to adjust the still prevailing surplus capacity to demand which eventually ceased to double every few years.

The formation of an Organisation of Petroleum Exporting

Countries (OPEC) in 1960, and its becoming an effective centre of decision making in the '70s, was virtually inevitable – nature abhors a vacuum. To maintain consistent cooperation of a disparate bunch of governments proves a daunting proposition – the successes of the organisation were so far confined to periods during which political rather than economic events created the perception of potential shortages. The structure became somewhat fragile once the biggest potential suppliers in the Gulf ceased to accept the role of swing producers – no cartel can survive unless the biggest members are prepared to make some sacrifices in order to make the weaker brethren toe the line.

One of the significant features of the current oil setup continues to be the wide gap between the low-cost crude oil available in massive amounts and the higher-cost production. This situation was aggravated by the almost accidental price explosions of the '70s which encouraged the exploration and development of non-OPEC oil and of alternative sources of energy. Once these sources were extant, and because countries and companies did not want to be at the mercy of OPEC, there was some degree of compulsion to keep them in being: the traditional policy of insulating the higher-cost supply by way of import levies and controls – previously sported by the USA for decades – did not fit into the prevailing free-trade policy climate and there now remains only an attitude designed to keep the price of low-cost oil at a level at which higher-cost supply can remain in the circuit.

Still there persists the paradox that much of the low-cost oil stays underground, whereas the higher-cost variety is being produced up to its technological limit. Consequently the OPEC members as a whole have been compelled to act as a swing producer, a role which its most potent countries have refused to play on their own.

V

The behaviour of the oil market so far has been determined by the consistent surplus of supply – actual and potential – over current and foreseeable demand.

The surplus capacity in tankers, refining and distribution,

the result of erstwhile anticipation of a rise in demand, which failed to materialize, has by now been considerably diminished. On the other hand the actual and the potential supply of crude oil dwarfs any expectations of demand down the road.

Consequently, failing some deliberate 'management' of supply there is, inevitably, being exerted strong and consistent pressure on the price level. There is, however, a prevailing line of thought, which expects to see a reversal of this trend in due course.

The fairly rapid exhaustion of the comparatively thin reserves of oil outside the Gulf would, within a few decades, still more concentrate the availability to what has been called the Real Oil, to give the Gulf states virtual control of supplies, which, in the hand of but a few agencies, could make it easy to control supply and thus the price level.

There is a lot in these thoughts, simply because of the stupendous concentration of huge low-cost reserves in this one singular point of the globe and because of the wide cost gap between it and almost all other competitive supply centres. Yet, the world *has* changed since the heady days of the '70s, when even the thought of an interruption of the supply of Gulf oil threw the world into a panic.

Oil, it must be at last admitted, is no longer the Last Frontier of contemporary life, the solitary vehicle and example of infinite progress. A number of events have brought on other sources of energy: natural gas, usually lying in strata deeper than oil, became available with the technique of deep drilling (as has much oil, following the development of the art of operating offshore) and transportation of gas has become a commercial proposition. Coal has become an international commodity, originating now in lower-cost areas. There is after all the vast opportunity of nuclear fission (to be followed, maybe, one day by fusion). Whereas the inherent technological danger has retarded this line of progress, the mounting concern about acid rain and greenhouse effect, identified as the result of hydrocarbon burning, has made ecological thinking begin to veer towards reconsidering the comparative merits of a safety-improved nuclear establishment. Finally, there is considerable scope in improving the way energy is being used.

It is true that each source of energy is, in one way or the other, finite, whereas potential demand may have no ceiling; yet a long time ago I have insisted that it was human ingenuity that had no limits.

All these elements would and will tend to curb any attempt to 'corner the market' and will force the would-be monopolists to constantly look over their shoulders, lest they create conditions for their own demise. Some recent events – such as the compression of costs offshore and, on another level, the astonishing progress made in Venezuela, are portents of things to come.

Consequently, the overriding relevance of the Gulf oil notwithstanding, we can envisage a future setup of greater variety than that to which we were accustomed hitherto.

VI

Until a decade or so ago the spot oil quotations, then in Europe called 'Rotterdam Prices', covered some fringe operations which, as it was then said 'indicated but exaggerated the prevailing trend'. From this marginal position the spot market has now moved into a more central situation. This is due to the Unit of Operation of the industry having been scaled down considerably.

As long as the mainstream oil companies were refining and marketing their own crude oil, there was but little call for smaller-scale short-term transactions. Now the picture is one of much greater variety, with even the biggest operators shopping around as buyers of feedstocks and with their having become themselves sellers of parcels of various sizes and locations.

Hence the role of traders as clearing houses as it were, has moved from the sidelines to the centre of the stage and an entirely new phenomenon – the straight speculator – has developed, almost overnight; the so-called Wall Street Refiner.

Thus oil has taken on to a great extent the features of classical commodity trading. This was brought about in the first instance by persistent and sometimes wide price fluctuations, lately the result of incomplete volume control by OPEC:

speculation thrives on the succession of ups and downs. On the other hand the uncertainty results in the emergence of *different* expectations and thus makes it possible for operators to hedge their risk, reducing thereby the perils involved in taking up any position at all.

Yet, they can hedge only within a limited time span, and thus the vagaries of the short-term market are of but limited use for long-term decision making, i.e. for the planning of investment.

If the signals emanating from the day-to-day market are not of much help to the planner, he has to fall back on projections of his own which, life being what it is, may turn out to be off centre. For enterprises to survive such possible errors of judgement, there is an overwhelming need for a strong built-in entrepreneurial equilibrium. This again presupposes a certain degree of size and diversification, which makes it possible to absorb untoward developments.

It is fascinating to realize that the slimmed down unit of operation to which I referred just now, and the emergence of a much wider range of operators, in the end may lead to a regrouping of a centre, formed by a small number of large enterprises; SURVIVAL OF THE FITTEST.

VII

Finally, one should analyze the degree to which the current setup lends itself to the creation of a climate in which the establishment and maintenance of reasonably stable conditions encourages the investment programmes on which future supplies inevitably do depend.

Ironically, OPEC the Big Bad Wolf of the '70s, whose overwhelming power appeared to be set to hold the world to ransom for ever, has painted itself into a corner, yet even most of the oil importing countries now find it to be a necessary part of the system. In fact the very existence of that club stops oil prices falling to well below $10 a barrel, a level to which it would be pushed, if the low-cost holders of reserves were driven to optimize by volume, once optimization by margin has proved to be impracticable. Whereas, the global revenue of, say, Saudi Arabia might remain immune for a while, the weaker OPEC members and much of non-OPEC production could be virtually wiped out, at least as far as development and new investment was concerned. It is worth noting that violent change of this nature may have disturbing macro-economic repercussions, especially since a total price collapse might be followed, for a while at least, by a comparable upsurge: hence for instance the Japanese, who benefit more than others from low energy import costs, carefully put their weight on the side of reasonable stability which fits in best into their concept of world markets. The cohesion of OPEC is most marked in extreme situations: either when they happen to be in control of the situation and both volume and price are optimal, or, at the other end of the scale, when prices have fallen to an extent that the need for hanging together, so as not to be hung separately, has become self evident.

In between the extremes, say, in the $15–18 range, and for 16 mbd output, the inherent centrifugal forces become virulent. Hence it has been said with some justification that the currently aimed at 'moderate' price target was the worst of all worlds for OPEC: too high to discourage non-OPEC competition, but not high enough for the members, due to their market share being hemmed in by competition, hence volume being unable to enhance their revenue.

The conclusion to be drawn from such analysis is that OPEC, as we now know it, is not sufficiently potent an agent to carry, on its own, the whole burden of keeping supply in line with demand, and that the rest of the world, in its own interest, has to lend a hand to achieve at least a modicum of continuity.

The apparent willingness of some non-OPEC exporters to coordinate their programmes with those of OPEC itself, although a breakthrough in this field is yet to come, is a straw in the wind and public opinion, even the narrow-minded circles now ruling for instance in the UK, are beginning to realise what is in its own long-term interest.

The prevailing intellectual climate is far from being conducive to positive action, but water tends to find its own level, and we might be about to enter a period of some sort of cooperation, implicit rather than explicit in the first instance.

There remains for me now to take leave from my readers and friends – I trust that there is a great deal of overlap between the two categories.

I do hope that the fundamental approach which I have propagated all through my professional life, and which has been briefly recounted just now, will continue to be a viable basis of industrial and governmental thinking. Be that as it may, it was great to be part of this industry for so long and to have been able to contribute to its intellectual setup.

To you all I wish the best of luck for your respective futures (you'll need it) and, perhaps more decorously, I should say: God Bless You All.

Epilogue

Paul Frankel's 'industrial philosophy', as he likes to call his system of thoughts on the oil industry, consists of simple but fundamental ideas. Simple ideas are not always obvious. The process that leads to their discovery is often arduous and roundabout. Once discovered and expressed they do not necessarily gain widespread acceptance, because the arrogant dismiss them as trite and the shallow-minded devalue them by confusing the simple with the superficial and the simplistic. Yet, the simple is often the seal which identifies the truth. *Simplex sigillum vero:* I have no doubt that Paul Frankel, the thoroughbred European, appreciates this Latin dictum. There is, in the same vein, an Arabic aphorism which refers more subtly to 'an elusive simplicity', the hallmark of great minds and distinguished writers; the simplicity and clarity of thought which, when encountered, seems so easy to emulate but which eludes everybody save the gifted few.

Paul Frankel's simple ideas are all summarized by himself in the preceding chapter. He calls it a Postscript but this is a misnomer, for what he has expressed so succinctly at the end was there, some forty years ago, at the beginning. It has 'imbued', as he would put it, all his writings and all his teaching.

The first idea is that market and government (or if you prefer economic forces and institutions) each have an essential role to play and need to co-exist. Frankel is not an ideologue (he is too much of a cynic to indulge in ideology) and does not believe that either the market or the state is always right. He has no time for those who say 'you cannot buck the market' and no time either for those who want to invest the government with all sorts of economic responsibilities. The simple truth is that we need both market and government; the difficult problem is to define the limits of their respective rules and to establish the rules for optimum interaction. I do not think

that Paul Frankel attempted this daunting task in any systematic way but he has often suggested, particularly in the
Topicals when commenting on some important event, how the
two protagonists should best play their respective roles. One
can but wish that the ideologues who today rule most of the
great Western democracies would recognize what Paul Frankel always saw with absolute clarity, that the search for 'purely monetary returns', a leitmotiv of the unfettered market
economy, 'leads to unacceptable social and political results'.
However, Frankel's main concern is oil and he argued forcibly
throughout his professional life that in oil (and by extension,
in energy) there are distinct roles for economic forces and for
governments. Oil matters are too serious to be left entirely to
the oil market.

The second basic idea is a fundamental proposition of oil
economics. The costs of finding and developing oil resources
are of a much higher order of magnitude than operating costs.
There are long lead-in times in oil investment which delay
supply responses to price increases; similarly the low operating costs weaken supply responses to a price fall. In short,
supply is inelastic in the short and medium term. Demand for
oil products, particularly in the transport sector, is also inelastic. For these reasons the oil industry is not *immediately* self-
adjusting. Adjustment is a long and clumsy process which
causes much destruction on its way. The lack of flexibility in
supply and demand parameters is an invitation to any powerful entity – be it a group of companies or a group of governments – to step forward and manage supply in order 'to bring
it in a workable relationship to demand'.

Paul Frankel saw merits in the management of supply as
operated in the 1930s and 1940s by the Seven Sisters in the
international market, and by prorationing agencies in the
USA. He realized, as early as 1957 and with remarkable
prescience, that the oil-exporting countries would want to step
in and regulate the market themselves. In a paper reproduced
here as Chapter 4, he stated: 'Finally the wheel may turn full
circle and the producer countries might one day call for an
international order as a means of securing for themselves an
"equitable" share in the markets.' He foresaw in 1957
OPEC's imminent arrival on the scene, but I am not sure he

trusted the oil-exporting countries to do as good a job at bringing supply and demand together at a reasonable price as the oil companies or the Texas Railroad Commission did. The man who believed so much in the importance of market regulation was not altogether reassured by the credentials of the new regulator. In October 1960, he commented: 'if the producer countries should succeed in finding a common de-nominator and become so impressed by the apparent power which they could have in forcing high prices on the world by restricting supplies that they should try to misuse this power, the repercussions could be serious on all sides.' This very early suspicion that OPEC may misuse its power inevitably col-oured his reactions to the 1973 events. He interpreted them as resulting from the exercise of monopoly power by sovereign states, which felt that they were no longer bound by the law of contract, preferring to determine their behaviour according to the law of changing circumstances. For Frankel the one thing that is worse than the disorder of the market is the disorder in international relations arising from countries' reneging on contracts.

My own interpretation of the 1973 events is different. The oil price increases were not due to the whims of capricious sovereign states; they were the inevitable consequence of a supply shortage caused by under-investment in the com-panies' concession areas in previous years. Put differently, the price increase of 1973 was the result of rigid price manage-ment in the previous decade. Falling oil prices stimulated demand faster than investment for additional supplies. OPEC's major sin in 1973 was not greed associated with monopoly power – as most of the world believed – but vanity. OPEC believed what everybody wanted it to believe – that it caused a major revolution and shock and found itself playing the dual (and rather contradictory) role, of hero and scapegoat.

Paul Frankel foresaw the problems which eventually beset OPEC. He wrote in 1956, rather prophetically: 'it would be in the interest of Middle East countries to establish an atmo-sphere in which the consumer countries would feel safe. Any doubt about their free access to Middle East oil on fair terms will encourage the investment in exploration and drilling elsewhere, even at high cost which normally would make such oil uneconomic. Once such oil is found in a consumer country the quantity is finally lost to the Middle East as an exporter because countries *will* make the best use of their own production once they have it, and will restrict imports.' Is not that, in a nutshell, the whole story of non-OPEC oil developments in the 1970s and 1980s? And he shared with many analysts the view that OPEC members would eventually succumb to the temptation of competing against each other for larger volumes. Optimizing by margin will give way to optimizing by volume whenever a shortage gives way to a glut. I suspect however that Frankel, although more far-sighted in these matters than some of his peers, did not expect OPEC to resist the temptation for as long as it did. Damaging competition began in late 1984; many thought, or rather hoped, that it would start in 1975!

But which type of market regulation would Frankel favour today? His advocacy for some mechanism that may bring supply and demand together without price shocks and his scepticism about OPEC's ability to provide the mechanism beg the question: who is the potential regulator? I think that Frankel would ideally favour the establishment of an international agreement, or rather a memorandum of understanding, between major consuming and producing countries. The hope is that such an arrangement would check the abuse of sovereign power by producing countries (in Frankel's view, an important cause of instability), but provide them, at the same time, with the benefits of orderly market developments. He did not mind too much the companies controlling the market in the 1930s and beyond because the companies are subject to the jurisdiction of governments who can limit their abuses; but states might be inclined to exceed their powers unless constrained by tight international supervision. I also suspect that Frankel, a student of the great Austrian economists, shared their distrust of governments. He went far enough in recognizing that governments have a role to play in relation to the market, but was not prepared to go further and allow them to claim exclusive economic sovereignty over their countries' natural resources. One cannot expect a maverick to denounce every tenet of the prevailing orthodoxy. Frankel, a son of the oil industry, believed in the ability of large concerns to get things done. He always argued that size, diversification and vertical integration were necessary for the efficient perform- ance of an oil company. And the logic of his own argument led him to predict very early on that oil-producing countries will, sooner or later, move downstream.

Frankel, the European, had the interests of Europe as he saw them at heart. I think that Europe for him is not the nations or their governments but the people, their industry and their prosperity. He witnessed a long period of economic growth fuelled by cheap oil, a period of fast industrial develop- ment led by a dynamic, yet stable, oil industry. He must have concluded that oil stability was important for the economic welfare of Europe: hence his teaching that the oil market must be regulated. And as a good Viennese economist he believed that the goal of all economic activity is either present or (through investment) future *consumption*. Thus, the regulation and the stability is for the consumers' ultimate benefit. In this perspective, the industry and the producing country are just instruments. They are of course entitled to a share of the economic rent, but merely as a reward for the job performed. Frankel probably did not fully appreciate that a producing country is also a society of consumers, and that some of the exporting nations of the Third World have immense require- ments for economic betterment and self-sustained develop- ment. It is not enough to worry about the European con- sumer, and from time to time about the oil-importing developing countries. The producing countries of the Third World have their problems too. His European sympathies led to charges, which he deeply resented, that he was only a spokesman for consuming countries' interests. The accusation of bias always hurts the honest intellectual who attempts to analyse econo- mic problems with a degree of objectivity. I understand his reaction and feelings, being myself sometimes accused of the opposite bias, a bias in favour of producing countries. My position on this issue is that no intellectual worth his or her salt is free of sympathies, emotions and value judgements. To deny this is akin to saying that man is soul without flesh. Yes, Frankel the European feels for the European consumers and understands their interests; yes, Mabro has empathy for Arab and non-Arab producers of the Third World and tries to understand their point of view. Call it bias if you wish; but without these biases the discovery of the truth would be impeded. We would miss much of the story if the consumers' point of view and the producers' point of view were not expressed, spelled out and explained by those who understand them with some empathy. The dangerous bias is that of propaganda. The dangerous bias is that which hides behind 'positive economics' and its spurious claims of objectivity. But the bias of those who declare their hand and apply a sympathetic mind to the understanding of a subject from within is a necessary ingredient of research. We can only assess the true merit of a case, be it that of consumers or producers, of governments or companies, of OPEC or the IEA, after hearing the best arguments put in its favour.

Paul Frankel's professional life coincided with the heyday of the petroleum industry. He worked, consulted and taught during an exciting period. He left his mark on several generations of men and women in companies and governments. He was and still is an authority. Did he have any peer during the five decades of his professional life? If I were to write a history of this period I would probably want to assess the contribution of five outside 'authorities' who had a significant influence on our understanding of oil matters. These are Frankel, Levy, Adelman, Penrose and Jablonski. Frankel, Levy and Jablonski have an unparalleled knowledge of the industry and great insights. Frankel used this talent to track, Levy to influence policy, and Jablonski to report and inform. Frankel is an economist with some understanding of politics, Levy a politician with an understanding of economics, and Jablonski an outstanding journalist with a sense of what is news in the context of history. Neither Adelman nor Penrose has Frankel's or Levy's understanding of the oil industry, that understanding from within which only accrues through daily exposure over decades of professional life. But as academic economists they forced the industry to look at its problems from a different angle, to question cherished myths, and to abandon comfortable tenets of the conventional wisdom. Penrose's main contribution was to show how fiscal regimes and transfer pricing explain the economic behaviour of the multinationals;

neither a simple nor an obvious idea when she developed it. Adelman's prowess was to apply the incisive tools of Marshallian analytics to the world petroleum market. Their contribution to oil economics was more substantial than Frankel's; their influence on the doers in the industry much less considerable.

None of the five is at once outstanding in economics, politics and industry matters. In fact the greater their talent in one area, the lesser their understanding of another; like Adelman, for example, a most distinguished economist with an appallingly naive approach to politics. None of them is truly international, but this is not too surprising. Nobody is really international; each of us is the child of a particular culture, always attached, even in exile, to some spiritual motherland. We might as well be grateful for that. Without this cultural diversity with its good values and irritating prejudices, these imbalances, this mosaic of talents, all research, criticism and seminal doubts would be stifled by some compelling authority, some all-embracing truth. It is good that great men and women are at once great and not so great.

The oil world is now changing almost beyond recognition. Stagnation has succeeded growth; atomistic competition has followed industrial concentration; short-term price speculation has replaced long-term price administration. Do these developments falsify Paul Frankel's simple but powerful message? Do they make the message irrelevant? I think, on the contrary, that they make it more actual and more pressing. The irony, however, is that the message of Paul Frankel, the European, the consumer-oriented Viennese economist, the son and admirer of the Anglo-Saxon and European oil industry, has today become a message for producing countries. Such is the Cunning of History; and if this probably disturbs Frankel the standard-bearer, it may at the same time amuse the cynic in him. Whether the oil-exporting countries will heed the message, I know not. Like Frankel, perhaps, I wish that they would, but like Frankel, certainly, I suspect that they will not.

Robert Mabro
September 1988

Index